Principles of SOIL CHEMISTRY

BOOKS IN SOILS AND THE ENVIRONMENT

edited by

E. A. Paul
Department of Plant and Soil Biology
University of California
Berkeley, California

Organic Chemicals in the Soil Environment, Volume 1, edited by C. A. I. Goring and J. W. Hamaker

Organic Chemicals in the Soil Environment, Volume 2, edited by C. A. I. Goring and J. W. Hamaker

Humic Substances in the Environment, by M. Schnitzer and S. U. Khan

Microbial Life in the Soil: An Introduction, by T. Hattori

Principles of Soil Chemistry, by Kim H. Tan

Also in the Series

Soil Biochemistry, Volume 1, edited by A. D. McLaren and G. H. Peterson

Soil Biochemistry, Volume 2, edited by A. D. McLaren and J. Skujiņš

Soil Biochemistry, Volume 3, edited by E. A. Paul and A. D. McLaren

Soil Biochemistry, Volume 4, edited by E. A. Paul and A. D. McLaren

Soil Biochemistry, Volume 5, edited by E. A. Paul and J. N. Ladd

Additional Volumes in Preparation

Principles of SOIL CHEMISTRY

KIM H. TAN

The University of Georgia
College of Agriculture
Athens, Georgia

MARCEL DEKKER, INC. New York and Basel

Library of Congress Cataloging in Publication Data

Tan, Kim Howard
Principles of soil chemistry.

(Books in soils and the environment)
Includes bibliographical references and indexes.
1. Soil chemistry. I. Title. II. Series.
S592.5.T36 1982 631.4'1 82-12997
ISBN 0-8247-1336-2

Marcel Dekker, Inc.
270 Madison Avenue, New York, New York 10016

Current printing (last digit):
10 9 8 7 6 5 4 3 2 1

Printed in the United States of America

Dedicated to the memory of

Dr. Ir. J. Van Schuylenborgh

The author's promotor and

deacon of the subfaculty of the

University of Amsterdam, Amsterdam,

The Netherlands

FOREWORD

Good college instruction consists primarily of a dedicated and knowledgeable instructor with organized information relevant to the subject. The rapid accumulation of knowledge and proliferation of interest in soil science by students not concentrating in agricultural studies could easily produce a teacher-student confrontation. A science course with little mention of soils or a provincial soils course without adequate science are always possibilities.

For a decade Dr. Tan has offered popular soils courses using available texts. However, students requested that a text be prepared to better complement the modern needs of students in understanding the complex nature of reactions in soils of importance in plant growth and crop production. Integrating pure science with the dynamics of soils is something Dr. Tan has accomplished in his research of fifteen years. He has now used the same procedure to write a solid textbook to teach a hard science to students of varied background. In the process he has balanced soil dynamics in a most interesting fashion. The text is scientific and the organization reinforcing, while the language is integrated and simple without being unnecessarily wordy. Not only has Dr. Tan written a book on soil chemistry, but he has written a book whose organization progresses in complexity from a conventional pure science to applied complex soil-plant relationships.

Elvis R. Beaty

The University of Georgia
Athens, Georgia

PREFACE

This book presents the adaptation of pure chemical science to the scientific study of soils and plants. The text provides comprehensive coverage of the fundamental topics in soil chemistry. Its unique approach, including a definite soils' flavor by the integration of organic and inorganic components in the dynamic processes in soils, important for continuation of life, is not currently available. In plain language, easy to comprehend by a wide range of scholars and students, but without sacrificing the scientific value, the book starts with a review of basic chemical principles and thermodynamics pertinent to the following topics on concepts and processes in and related to the soil solution: colloidal organic and inorganic components, their modern classification, and reactions and interactions affecting changes in the behavior of soils and plant growth. The book tells you everything that you want to know about humic acids. Separate chapters are included on the use of x-ray diffraction, infrared analysis, differential thermal analysis (DTA), and other methods for the identification of organic and inorganic soil constituents. Examples are given in interpretation of results using tables provided on diagnostic d spacings in x-ray analysis, wavenumbers in infrared spectroscopy, and peak temperatures in DTA. Crystal chemistry of inorganic compounds and surface chemistry of inorganic and organic colloids are explained in simple terms, showing their significance in the control of the many complex reactions in nature. The traditional adsorption, cation and anion exchange, soil acidity, and salinity theories are presented together with the current concepts. In the final two chapters, the principles of soil chemistry are applied in soil formation and in soil-organic matter interaction. Although the purpose of the book is to fill a need in soil science, that is, by bridging pure chemistry and soil science, the volume is equally useful in explaining the soil as a basic entity for related disciplines in agriculture and other sciences, for example, crop and plant sciences, irrigation, forestry, conservation, plant physiology, ecology, microbiology, geology, geochemistry, physics, chemistry, and botany.

Special recognition goes to Dr. Elvis R. Beaty, professor of agronomy, University of Georgia, for editing the book. My thanks are also due to Dr. Ralph A. Leonard, research leader, Southeast Watershed Research Program, USDA,-SEA, -AR, and to Dr. Robert A. McCreery, associate professor of agronomy, University of Georgia, for reading the manuscript for correct English usage and scientific value. Appreciation is extended to Dr. J.B. Jones, former director, Soil Testing Laboratory, Cooperative Extension Service, and former division chairman, Department of Horticulture, University of Georgia, for his valuable criticism, and to the unnamed people, who have assisted in the development of the book. Thanks are also extended to the various publishers, scientific societies, and fellow scientists, who gave permission to reproduce figures, photographs, and diagrams. Finally, the author wants to thank his wife, Yelli, and his son, Budi, who always stood by with great enthusiasm and a lot of encouragement.

Kim H. Tan

CONTENTS

Principles of SOIL CHEMISTRY

1

REVIEW OF BASIC CHEMICAL PRINCIPLES

1.1 ATOM AND ATOMIC STRUCTURE

An atom is the smallest particle of an element that can enter into a chemical combination. Atoms of the same elements are similar in composition, but one element differs from the other in size, position, and movement of its atoms. An element is a substance composed of atoms with the same atomic number, or nuclear charge. In solid matter, the atoms vibrate within the confines of very small spaces, whereas in gas the atoms exhibit a considerable range of movement.

The concept of atoms being the smallest particles of matter was first postulated by Democritus or Leucippus in approximately 425 B.C., but it was not before Dalton's atomic theory was formulated in the first decade of the nineteenth century that this idea became scientifically established. Since then Crookes, Thomson, and others, working on the conduction of electricity in rarefied gases, made revisions in the theory above and concluded that the atom was composed of still smaller particles. The structure of the atom became a subject of research interest, and by the end of the nineteenth century it became known that the atom had the following components:

1. Electrons, small negatively charged components of atoms of all substances
2. Protons, positively charged particles of much greater mass than electrons

With the advancement of science in the twentieth century, it became clear that atoms also contain neutrons. The neutrons have a mass number of 1, but have zero (0) charge. Less fundamental particles were also detected, the positrons. Positrons are particles with the mass of an electron and the charge of a proton. The fundamental particles of the atom recognized today are (1) electrons, (2) protons, and (3) neutrons.

Protons and neutrons are located in a small central portion of the atom called the *nucleus*. The nucleus is of high specific weight and contains most of the mass of the entire atom. The various groups of electrons are placed in concentric shells around the nucleus. The shells may be composed again of subshells or cells. Neglecting the presence of subshells, the shells may contain one electron, as is the case with a hydrogen atom, or two electrons, as is the case with a helium atom, or more (Figure 1.1). The first shell adjacent to the nucleus is called the *K shell*, while the shell next to it is designated as the *L shell*, and so on. The largest atom is the uranium atom (^{238}U) with 92 electrons distributed around the nucleus in K, L, M, N, O, P, and Q shells. The diameter of the nucleus is between 1×10^{-13} and 1×10^{-12} cm. While the nucleus carries an integral number of positive charges, or integral number of protons, each of 1.6×10^{-19} C, each electron carries one negative charge of 1.6×10^{-19} C.

The electrons in the inner shells are tightly bound to the nucleus. This inner structure can be altered by high-energy particles (α rays, x-rays). With most atoms, it is the arrangement of energy in the outer shells that undergoes changes during chemical reactions. These outer shell electrons are responsible for the chemical properties of the element. During these changes, the role of the nucleus is usually a passive one. The hydrogen atom is perhaps an exception since it has only one shell and one bare proton.

An atom which loses one or more electrons from the outer shell is called a *cation* (Faraday), since such an atom assumes a net positive charge. When an atom has excess electrons, not balanced by the

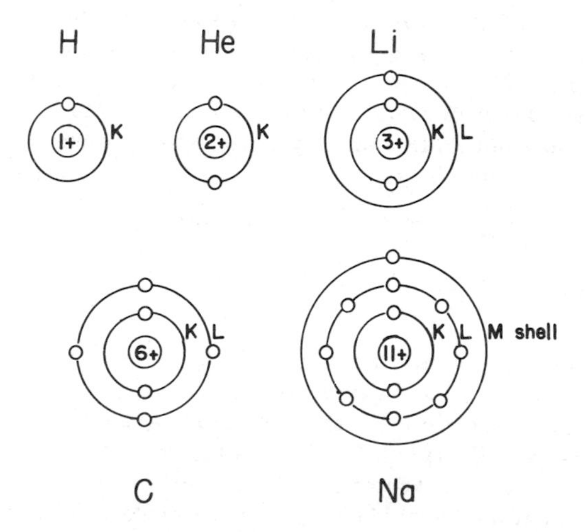

Figure 1.1 Atomic structure showing the K, L, and M electron shells.

positive charges of the nucleus, it assumes a net negative charge and is called an *anion*.

1.2 ATOMIC MASSES AND WEIGHTS

The atomic mass number is the sum of the masses of the protons and neutrons. Masses of electrons are neglected since they are very small and are considered to form an insignificant portion of the total mass of an atom. The atomic mass can be expressed on a chemical or a physical scale. In general, masses on the chemical scale are smaller than a weighted average physical mass by a factor of 0.99973. On a chemical scale the element oxygen is arbitrarily assigned the mass 16, nitrogen the mass 14, hydrogen the mass 1, and so forth. For a pure isotope the atomic weight gives the total number of nucleons (protons + neutrons).

Strictly speaking, atomic weights are not weights at all. They are reference numbers which indicate the relative weights of the different kinds of atoms, and no reference is made to absolute weights. Hydrogen was originally assigned a relative weight of 1, since it was considered the fundamental particle and the lightest of all atoms. The heaviest of any of the naturally occurring atoms is uranium with an atomic weight of 238. When we say that the element oxygen has an atomic weight of 16, we simply indicate that the oxygen atom is 16 times heavier than the hydrogen atom. Therefore, no weight units have been assigned to the numbers.

Atomic weights apply only to elements. Compounds have molecular weights, which equal the sum of the atomic weights of all elements making up the compounds.

The initial selection of hydrogen as a standard base for comparative assessment of other atomic weights was later changed in favor of oxygen, which is assigned an arbitrary mass of 16.0000. The atomic weight of hydrogen changed consequently to 1.0080. To comply with suggestions from atomic physicists, this oxygen base was revised again in 1961, and atomic weights (AW) are currently based on the assigned relative mass of ^{12}C (AW 12.0000). However, the consequent changes in atomic weights of the other elements are very small.

1.3 AVOGADRO'S NUMBER

The number of atoms in 1 gram atomic weight of any element is

$$6 \times 10^{23}$$

This number is known as the *Avogadro's number* (See Table 1.1).

How much does one atom of hydrogen weigh? Since 1 gram atom of hydrogen weighs 1 gm, and since 1 gram atom of hydrogen contains 6×10^{23} atoms, then one atom of hydrogen will weigh $1/6 \times 10^{23} = 1.67 \times 10^{-24}$ g, an extremely small number. Even the actual weight of an uranium atom is inconceivably small. Yet, it is 238 times heavier than is a hydrogen atom.

1.4 ATOMIC NUMBER

The *atomic number* is the number of protons, and hence the nuclear charge (symbol Z), in the atomic nucleus. Then the number of neutrons equals the difference between the atomic weight and the atomic number Z.

1.5 ATOMIC ORBITALS

Atomic orbitals are functions which define the spatial behavior of electrons of given energy levels in a particular atom. The number of orbital electrons is equal to the number of protons in the nucleus, so that the atom as a whole has a net charge of zero.

1.6 ATOMIC RADIUS

The electron density around an isolated atom extends to infinity. However, if one is referring to the size of an atom in a molecule or a crystal, then one may define the *atomic radius* as the closest distance of approach of a probe to the nucleus (Table 1.2).

As we will see later on, the size of elements will play an important role in many soil chemical reactions. Cation exchange reactions, hydration, double layers, and problems with potentials, dispersion, and flocculation of colloids, and so on, are all affected by the size of ions in the reactions.

1.7 VALENCE

The *valence* of an atom or element is that property which is measured by the number of atoms of hydrogen (or its equivalent) that can be held in combination by one atom of that element if negative, or that can be displaced by one atom of the element if it is positive. In simple terms, the valence is a measure of the combining capacity of atoms. Atoms with the smallest combining capacity are considered to have a valence of 1. Valences are whole (integral) numbers and

Table 1.1 Absolute Weights of One Atom (or One Molecule) of Selected Elements

Substance	Mole	AW or MW	Grams	Number of particles or atoms	Weight of one atom
H^+ ion	1	1	1	6×10^{23}	1.67×10^{-24}
Carbon	1	12	12	6×10^{23}	2.0×10^{-23}
Na	1	23	23	6×10^{23}	3.83×10^{-23}
K	1	39	39	6×10^{23}	6.50×10^{-23}
Ca	1	40	40	6×10^{23}	6.67×10^{-23}
NaCl	1	58	58	6×10^{23}	9.67×10^{-23}
KCl	1	74	74	6×10^{23}	1.23×10^{-22}
$CaCO_3$	1	100	100	6×10^{23}	1.67×10^{-22}
$C_6H_{12}O_6$ (glucose)	1	180	180	6×10^{23}	3.00×10^{-22}

Table 1.2 Ionic Radii, Atomic Numbers, and Atomic Weights of Selected Elements

	Radius (Å)[a]			
Ion	Crystalline	Hydrated	Atomic number	Atomic weight
Si^{4+}	0.42	—	14	28.09
Al^{3+}	0.51	9	13	26.98
Fe^{2+}	0.74	—	26	55.84
Fe^{3+}	0.64	9	26	55.84
Ca^{2+}	0.99	4.3	20	40.08
Mg^{2+}	0.66	4.2	12	24.32
Ba^{2+}	1.35	4	56	137.36
Li^{+}	0.60	3.8	3	6.94
Na^{+}	0.98	3.6	11	22.99
K^{+}	1.33	3.3	19	39.10
Rb^{+}	1.48	3.3	37	85.48

[a] 1 Å (angstrom) = 1×10^{-8} cm or 1 Å = 0.1 nm.
Source: Weast (1972) and Gast (1977).

correspond to the number of valence electrons the atom carries. Valence electrons are electrons which are gained, lost, or shared in chemical reactions.

1.8 EQUIVALENT WEIGHT

The *equivalent weight* (EW) of an atom or ion is defined as

$$\frac{\text{Atomic (formula) weight}}{\text{Valence}}$$

Elements entering into a reaction always do so in amounts proportional to their equivalent weights. Examples for the determination of equivalent weights are shown below:

1. With monovalent ions, 1 equivalent equals 1 mole.
2. With polyvalent ions, the equivalent weight is variable and depends on the reaction:

$$H_3PO_4 \rightarrow H^+ + H_2PO_4^- \qquad 1\ Eq = 1\ mol$$

$$H_3PO_4 \rightarrow 2H^+ + HPO_4^{-2} \qquad 1\ Eq = \frac{1}{2}\ mol$$

$$H_3PO_4 \rightarrow 3H^+ + PO_4^{-3} \qquad 1\ Eq = \frac{1}{3}\ mol$$

3. In precipitation and complex reactions, the relationship between the equivalent weight and numbers of moles can be read directly from the reaction. In volumetric analysis by which cyanide is titrated with silver according to the Mohr method, the reaction occurs as follows:

$$Ag^+ + CN^- \rightarrow AgCN$$

In this case the equivalent equals 1 mol. If the Liebig method is used in the titration of cyanide with silver, the end point of titration is reached when the following reaction has occurred:

$$Ag^+ + 2CN^- \rightarrow Ag(CN)_2^-$$

Consequently the equivalent of cyanide equals 2 mol.

4. Oxidation-reduction reactions.

In these reactions, the equivalence of a substance is by definition that part of a mole which in its reaction corresponds to the removal of ½ gram atom of oxygen, or the combination with 1 gram atom of hydrogen or any other univalent element. One way to find the equivalent is to determine in the reaction the change in oxidation state of the element, as shown in the following examples:

1. In the titration of ferrous into ferric iron using an oxidation agent, the state of oxidation of iron changes from 2 to 3:

$$Fe^{2+} \rightarrow Fe^{3+}$$

Therefore, the equivalent of ferrous iron equals 1 mol.

2. On the other hand, in the oxidation of metallic iron to ferric iron, the change in oxidation state is from 0 to 3:

$$Fe \rightarrow Fe^{3+}$$

The equivalent of metallic iron, therefore, equals 1/3 mol.

3. In volumetric analysis, where permanganate is used as an oxidizing agent in acid medium, permanganate ion is reduced into manganous ions:

$$MnO_4^- \rightarrow Mn^{2+}$$

or

$$Mn^{7+} \rightarrow Mn^{2+}$$

The change in oxidation state of Mn is from 7 to 2, which means a change in five units. The equivalent weight of permanganate is therefore 1/5 mol. If used in neutral medium, the permanganate is reduced to MnO_2:

$$MnO_4^- \rightarrow MnO_2$$

or

$$Mn^{7+} \rightarrow Mn^{4+}$$

The state of oxidation of permanganate changes with three units. The equivalent then is 1/3 mol.

As indicated in these examples, it is often not necessary to write the entire balanced equation of the reaction in order to find the equivalent weight. It is sufficient to write down only the change in oxidation state.

Another way to determine the equivalent weight (EW) follows. Instead of using the state of oxidation, the equivalent weight can also be found by the number of electrons transferred in the oxidation-reduction reactions. The following reactions serve as examples:

$$\text{Equivalent weight} = \frac{\text{molecular weight}}{\text{number of electrons lost or gained}}$$

$$Fe^{2+} \rightarrow Fe^{3+} + e^- \qquad EW = \frac{Fe}{1}$$

$$Sn^{2+} \rightarrow Sn^{4+} + 2e^- \qquad EW = \frac{Sn}{2}$$

$$Fe(CN)_6^{4-} \rightarrow Fe(CN)_6^{3-} + e^- \qquad EW = \frac{Fe(CN)_6}{1}$$

$$As^{3+} \rightarrow As^{5+} + 2e^- \qquad EW = \frac{As}{2}$$

$$MnO_4^- + 8H^+ + 5e^- \rightarrow Mn^{2+} + 4H_2O \qquad EW = \frac{MnO_4}{5}$$

$$MnO_4^- + 4H^+ + 3e^- \rightarrow MnO_2 + 2H_2O \qquad EW = \frac{MnO_4}{3}$$

$$Cr_2O_7^{2-} + 14H^+ + 6e^- \rightarrow 2Cr^{3+} + 7H_2O \qquad EW = \frac{Cr_2O_7}{6}$$

$$VO_4^{3-} + 6H^+ + e^- \rightarrow VO^{2+} + 3H_2O \qquad EW = \frac{VO_4}{1}$$

1.9 NORMALITY

The number of equivalents of a substance dissolved in 1 liter of a solution determines the *normality* ($\underline{N}$) of the solution. If one equivalent is present in 1 liter of solution, the solution is 1 normal. The symbol $\underline{N}$ (normality) is usually underlined to distinguish it from the symbol N, the nitrogen element.

1.10 ISOTOPES

Isotopes are defined as the same elements with a similar atomic number, but with a differing mass number. It became apparent that not all atoms of the same element had the same atomic weight. Mass spectrographic analysis yielded evidence that oxygen could be separated into three types of oxygen with mass numbers of 16, 17, and 18, respectively. These types of oxygen are called isotopes of the oxygen element. They are different forms of the same element. The atoms of the isotopes have the same number of protons, but a different number of neutrons. As stated above, oxygen has three isotopes: ^{16}O, ^{17}O, and ^{18}O. However, 99.76% of all oxygen is in the form of ^{16}O. Hydrogen is known also to have three isotopes, $_1^1H$, $_1^2H$ (called *deuterium*), and $_1^3H$ (called *tritium*). Deuterium and tritium are rare.

1.11 RADIOACTIVITY

Radioactivity involves a spontaneous disintegration of certain types of elements or atoms to form other elements or atoms. Rutherford made the discovery that if atoms were able to eject α particles, then these atoms were transformed into atoms with lower atomic weights. Ejection of α particles means that two neutrons and two protons are emitted as a single particle. The remaining part of the atom is a new element. Among the many types of radioactive elements, perhaps uranium is the best known. By emission of an α particle, the uranium nucleus loses two positive charges, and its atomic number drops from 92 to 90. Since the ejection of an α particle also means a loss of two neutrons, the mass number of the new element decreases from 238 to

234. This new element is called *thorium*. In turn, thorium may emit electrons (β particles), which is the result of a neutron decaying into a proton and an electron. A new element is then formed, called *proactinum*, with an atomic number of 91.

The emission of either electrons or α particles from the nucleus of a decaying atom is a natural process. Before radioactivity stops, or the nucleus becomes stable, $^{238}_{92}U$ will continuously exhibit radiation by emission of α particles. The final product is usually lead 206 ($^{206}_{82}Pb$).

1.12 HALF-LIFE OF RADIOACTIVE MATERIAL AND CARBON DATING

Half-life is a measure of the rate of decay of radioactive material. It is the length of time during which the material loses one-half of its radioactivity.

Radioactive elements emit a definite number of particles per second. This rate of decay is dependent on the amount of radioactive material, and the rate of emission will, therefore, decrease as the amount of decaying atoms becomes gradually smaller and smaller as emission of particles progresses. For many of the radioactive materials the time for complete decay is practically infinitely long, and it is very difficult, if not impossible, to determine the life span of these radioactive materials. However, the half-life span is relatively easily determined. The half-life of a radioactive element is a characteristic feature that can not be altered by physical or chemical means. Examples of characteristic half-lives of selected elements are shown below (Weast, 1972):

Element	Half-Life		Element	Half-Life	
U^{238}	4.5×10^9	years	Co^{57}	270	days
U^{234}	2.5×10^5	years	Fe^{55}	2.6	years
Ra^{226}	1620	years	Mn^{53}	2.0×10^6	years
Pb^{207}	Stable		Ca^{45}	165	days
Pb^{206}	Stable		K^{42}	12.4	hr
Ba^{131}	12	days	Cl^{36}	3.1×10^5	years

Element	Half-Life		Element	Half-Life	
Zn^{65}	243.6	days	S^{35}	88	days
Cu^{64}	12.9	hr	P^{32}	14	days
Si^{31}	2.62	hr	Na^{22}	2.6	years
Mg^{28}	21	hr	C^{14}	5730	years
Al^{26}	7.4×10^5	years			

Since the rate of decay for a given amount of radioactive material is constant, and since the final product is a stable Pb isotope, the ratio Pb concentration/^{238}U concentration is considered a measure for the minimum age of the sample. The principle above is applied in the analysis for determining the age of geologic and anthropologic material with ^{14}C. This analysis is called *carbon dating*. Carbon 14, ^{14}C, is a radioactive isotope, that has been formed by cosmic rays from outer space hitting nitrogen atoms in the atmosphere. The latter is then transmuted into ^{14}C, which decays into ^{12}C. An equilibrium usually exists in nature between the rate of ^{14}C formation and its rate of decay into ^{12}C. During their growth, the organism in nature obtains their carbon from the atmosphere directly or indirectly, and both ^{12}C and ^{14}C are absorbed in proportion to the equilibrium concentration existing in the air. As soon as these organisms die, the uptake of ^{12}C and ^{14}C ceases, but radioactive decay of ^{14}C into ^{12}C continues. By measuring the amount of ^{14}C in the material, the length of time since death can be estimated. Fresh or living plant material contains more ^{14}C than decomposed or dead residue. Since the half-life of ^{14}C is only 5370 years, carbon dating is limited in its determination of age to approximately 10,000 to 15,000 years. For older materials, other radioactive methods are available.

2

ELECTROCHEMICAL CELLS AND CHEMICAL POTENTIALS

2.1 ELECTROCHEMICAL CELLS AND ELECTRODE POTENTIAL

The soil system is the reservoir of most plant nutrients and also contains the active surfaces, which determine the concentration of ions in the soil solution. Ion movement, accumulation, availability of elements and uptake by plants, changes in element oxidation and reduction state, and many other chemical reactions in soils are reactions which to a certain extent show some resemblance to those occurring in an electrochemical cell. In pure chemistry two types of electrochemical cells are distinguished: (1) galvanic or voltaic cells and (2) electrolytic cells. A galvanic cell consists of two electrodes and one or more solutions (two half-cells). It is capable of spontaneously converting chemical energy from the solutions into electrical energy and supplying this energy to an external source. The automobile battery is an example of this kind of chemical cell. When one of the chemical components responsible for the reaction is depleted, the cell is considered dead. In an electrolytic cell, electrical energy is supplied from an external source. Electrochemical changes are produced at the electrode-solution interfaces (see Figure 2.1), and concentration changes are developed in the bulk of the system. If the external current is turned off, the system tends to produce current in the opposite direction. The lowest external electromotive force (emf) that must be applied in order to bring about the continuous separation of cations and anions (electrolysis) is called the *decomposition voltage* or the *back emf*. At the exact point where the galvanic emf is opposed by an equal applied emf, no current flows in either direction. In this static condition, the potential generated at the interface of an indicator electrode reflects the composition of the solution phase.

Now that we know what electrochemical cells are, let us assume that we have a pure Cu electrode dipping into a solution containing

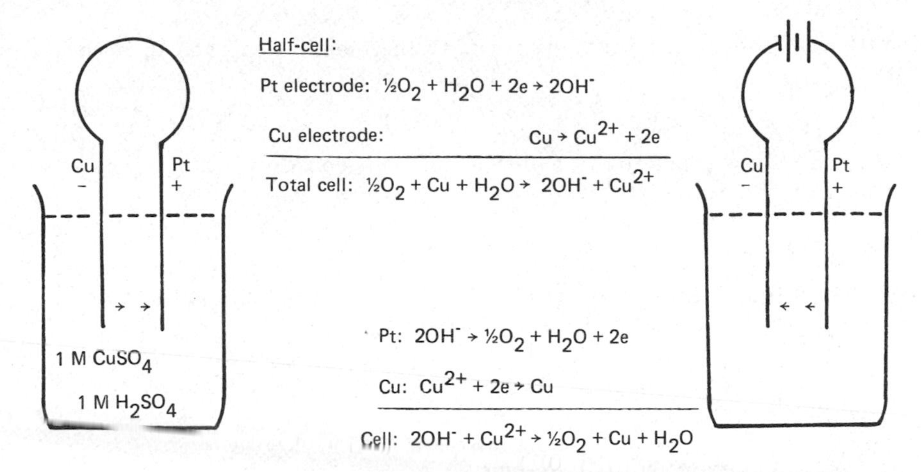

Figure 2.1 Schematic diagram of half-cell and total cell reactions in an electrolytic cell.

cupric ions. At the interface, Cu tries to dissolve from the metal, increasing the Cu concentration in the solution. Therefore, a potential difference develops on the surface of the Cu electrode, between the metal and the solution. This potential difference is formulated by the Nernst equation

$$E = \frac{RT}{nF} \ln \frac{K}{M^n} \tag{2.1}$$

where

E = potential difference
R = gas constant
n = valence of the ion
F = Faraday constant
T = absolute temperature
K = impulse of metal to dissolve
M = ion activity

However, it is not possible to measure the potential of a single electrode. It is only possible to measure the potential of one electrode relative to another, the reference electrode. A standard reference electrode is the hydrogen electrode (Pt electrode dipping in a solution at unit activity of hydrogen ions). If the two electrodes are connected and an external emf is applied, electrons will flow through the system. Copper will be deposited on the Cu electrode:

$$Cu^{2+}\ (aq) + 2e^- \rightleftharpoons Cu\ (c)$$

At the hydrogen electrode, hydrogen will release electrons to become H^+:

$$H_2(g) \rightleftharpoons 2H^+(aq) + 2e^-$$

Either of these reactions is called a *half-cell reaction*. The overall reaction is

$$Cu^{2+}(aq) + H_2(g) \rightleftharpoons Cu(c) + 2H^+(aq)$$

The potential of the Cu electrode (half-cell) measured against the standard half-cell electrode is

$$E_h = E^o + \frac{RT}{nF} \ln \frac{Cu^{2+}}{Cu} \qquad (2.2)$$

At 25°C, this equation can be written as

$$E_h = E^o + \frac{0.0592}{n} \log \frac{Cu^{2+}}{Cu}$$

E_h is called the *electrode potential*, whereas E^o is the standard potential of the cell in which the reactants and products have unit activity. The electrode potential has been defined by the International Union of Pure and Applied Chemistry (IUPAC, 1960) as the electron availability or the electrochemical potential of the electron at equilibrium. The formula of E_h can be generalized as follows:

$$E_h = E^o - \frac{RT}{nF} \ln \frac{\text{reduction}}{\text{oxidation}} \qquad (2.3)$$

The standard potential of a cell when the other electrode is a standard hydrogen electrode is by definition called the *standard electrode potential.*

Depending on the way the reaction is written, electrode potentials are oxidation or reduction potentials. If the reaction is written as a reduction reaction,

$$Cu^{2+}(aq) + 2e^- \rightleftharpoons Cu(c)$$

the standard potential is called a *reduction potential*. Both electrode and reduction potentials of Cu are then positive in sign. If, however, the reaction is written as an oxidation reaction:

$$Cu_{(c)} \rightleftharpoons Cu^{2+}(aq) + 2e^-$$

the standard potential equals to an oxidation potential and is negative in sign. The standard potential of the hydrogen electrode is by convention zero. A number of selected potentials are given in Table 2.1

Table 2.1 Selected Electrode Potentials of Half-Cell Reactions

Half-cell reaction[a]	Electrode potential (volts)
$3N_2 + 2H^+ + 2e^- \rightleftharpoons 2NH_3$	-3.100
$Li^+ + e^- \rightleftharpoons Li$	-3.045
$K^+ + e^- \rightleftharpoons K$	-2.924
$Ba^{2+} + 2e^- \rightleftharpoons Ba$	-2.900
$Na^+ + e^- \rightleftharpoons Na$	-2.710
$Mg^{2+} + 2e^- \rightleftharpoons Mg$	-2.375
$ZnO^{2-} + 2H_2O + 2e^- \rightleftharpoons Zn + 4OH^-$	-1.216
$Mn^{2+} + 2e^- \rightleftharpoons Mn$	-1.029
$Fe^{2+} + 2e^- \rightleftharpoons Fe$	-0.409
$Fe^{3+} + 3e^- \rightleftharpoons Fe$	-0.036
$2H^+ + 2e^- \rightleftharpoons H_2$	0.000
$Cu^{2+} + e^- \rightleftharpoons Cu^+$	+0.158
$AgCl + e^- \rightleftharpoons Ag + Cl^-$ (silver-silver chloride electrode)	+0.222
Calomel electrode, 1N KCl	+0.280
$Cu^{2+} + 2e^- \rightleftharpoons Cu$	+0.340
$Fe^{3+} + e^- \rightleftharpoons Fe^{2+}$	+0.770
$Hg_2^{2+} + 2e^- \rightleftharpoons 2Hg$	+0.789
$Ag^+ + e^- \rightleftharpoons Ag$	+0.799
$MnO2 + 4H^+ + 2e^- \rightleftharpoons Mn^2+ + 2H_2O$	+1.208
$O_2 + 4H^+ + 4e^- \rightleftharpoons 2H_2O$	+1.229

Table 2.1 (continued)

Half-cell reaction[a]	Electrode potential (volts)
$Cr_2O_7^{2-} + 14H^+ + 6e^- \rightleftharpoons 2Cr^{3+} + 7H_2O$	+1.330
$MnO_4^- + 8H^+ + 5e^- \rightleftharpoons Mn^{2+} + 4H_2O$	+1.510
$H_2O_2 + 2H^+ + 2e^- \rightleftharpoons 2H_2O$	+1.770

[a]In accordance to IUPAC conventions, the reactions above are written as reduction reactions. The signs of electrode potentials may perhaps be opposite in signs in other books if written as oxidation reactions.
Source: Weast, 1972.

The difference of two standard potentials gives the standard potential of the desired reaction.

Electron availability or redox potential is an indication of the oxidation-reduction status of soils. It affects the oxidation states of H, C, N, O, S, Fe, Mn, Cu, and many other elements, and as such controls solubility and availability of many nutrient elements to plants. But for the oxidation-reduction limits imposed by the stability of water, this list would include all the elements of the periodic table. The limit of oxidizing conditions in aqueous systems is the oxidation of water to molecular oxygen. On the other hand, the limit of reducing conditions is the reduction of the hydrogen ion to molecular hydrogen. In natural systems, redox potentials have often been treated as equilibrium potentials. At equilibrium, a mixture of redox couples reacts until the net donation and acceptance of electrons is zero. The electron's escaping tendency or the E_h of each redox couple is the same, and their reduction/oxidation ratios have adjusted to that defined by the Nernst equation.

However, natural soil systems rarely reach oxidation-reduction equilibrium, because of the continuous addition of electron donors, i.e., oxidizable organic compounds. The oxidation of these compounds is often very slow, even, when the major electron acceptor, oxygen, is available. Redox conditions in living systems are nonhomogeneous, even within a single cell. In photosynthesis, for example, water is split into a strongly reducing form of hydrogen and the strong oxidizing agent oxygen, within a chloroplast.

2.2 CHEMICAL POTENTIALS AND THEIR APPLICATION IN ION UPTAKE BY PLANTS

To each chemical species in a reaction mixture, a certain amount of (free) energy can be ascribed. This amount of energy, expressed per unit amount of ion species, is called *chemical potential*. This entity, indicated by the symbol μ, depends on the pressure P, the temperature T, the chemical nature of the ion species, and on its mixing ratio with other species. For ideal solutions the chemical potential can be formulated as follows:

$$\mu = \mu^\circ + RT \ln m \tag{2.4}$$

where

μ = chemical potential of an ion species
μ° = chemical potential of the ion species at standard state
m = moles of ion species in the mixture
R = gas constant
T = absolute temperature

For nonideal solutions, Eq. (2.4) above changes to

$$\mu = \mu^\circ + RT \ln a \tag{2.5}$$

in which a is the activity of ion species. At infinite dilution (or very dilute solutions) Eq. (2.5) can be changed into

$$\mu = \mu^\circ + RT \ln c \tag{2.6}$$

in which c denotes concentration of the ion species. Since μ° is considered a constant, the chemical potential can be written as $\mu = RT \ln m$ (ideal), or $\mu = RT \ln a$ (nonideal) plus a constant.

As stated earlier the chemical potential indicates the state of potential energy of the chemical species or component in soils, and its formulation shows some relationship with the Nernst potentials. It is independent of external force fields, such as gravity or centrifugal forces, and the chemical potential will remain constant at equilibrium conditions, e.g., constant concentrations, temperature, and pressure. Differences in chemical potential of a species at various locations in the soil tend to induce spontaneous movement of the species in the direction of points with lower potentials.

The chemical potential as formulated above can be applied to describing ion transport, movement, and adsorption in soils, and is very useful in prediction of ion uptake by plant roots and of ion transport from cell to cell. The selective uptake of chemical compounds by the root system is an outstanding process of living systems. Cell compartments in the plant tissue are separated by biological membranes representing barriers for chemical compounds. The transport through these barriers and processes mediating this transport can be studied and described using chemical potentials. Although not clearly understood, it is currently accepted that plant cells behaving as biological membranes have pores acting as sieves, favoring penetration of small particles only. This kind of passive transport (diffusion, mass flow) of small hydrophillic particles obeys physical and chemical laws. The net ion flux in either direction will stop as soon as a state of equilibrium is reached. At equilibrium the system conforms to the Donnan principle. With compounds that are not electrically charged, e.g., sucrose, the equilibrium is attained when equal sucrose concentrations exist on either side of the membrane. However, with ions possessing electrical charges, the electrochemical potentials must be equal on both sides of the membrane:

$$\mu_i + zF\psi_i = \mu_o + zF\psi_o \tag{2.7}$$

where

i = inside cell
o = outside cell
ψ_i = electrical potential at innerside of membrane
ψ_o = electrical potential at outerside of membrane
μ = chemical potential of element species
z = valence
F = Faraday constant

The chemical potential was defined earlier as $\mu = RT \ln a$; therefore substituting $RT \ln a$ for μ, Eq. (2.7) becomes

$$(RT \ln a)_i + zF\psi_i = (RT \ln a)_o + zF\psi_o$$

$$zF\psi_o - zF\psi_i = (RT \ln a)_i - (RT \ln a)_o$$

$$\psi_o - \psi_i = E = \frac{RT}{zF} \ln \frac{a_i}{a_o} \tag{2.8}$$

$\psi_o - \psi_i$ is called the *membrane* (or *Donnan*) *potential*, and by changing it into symbol E, Eq. (2.8) conforms to the Nernst equation. Because $\psi_o - \psi_i$ is a positive figure, we have

$a_i > a_o$ for cations

$a_i < a_o$ for anions, z is negative

At equilibrium, the chemical potential of cations also equals the chemical potential of anions. This is necessary to maintain electroneutrality. Therefore

$$E_c = \frac{RT}{zF} \ln \left(\frac{a_i}{a_o} \right)_c \quad \text{equals} \quad E_{an} = \frac{RT}{zF} \ln \left(\frac{a_o}{a_i} \right)_{an}$$

where

c = cations
an = anions

Consequently

$$\left(\frac{a_i}{a_o} \right)_c = \left(\frac{a_o}{a_i} \right)_{an} = \text{constant} \qquad (2.9)$$

The latter means that the ion product on either side of the membrane is constant. If KCl or $CaCl_2$ is present in the system, then according to the Donnan equilibrium principle the following is valid:

$$(K^+)(Cl^-)_{inside} = (K^+)(Cl^-)_{outside}$$

or

$$(\sqrt{Ca^{2+}})(Cl^-)_{inside} = (\sqrt{Ca^{2+}})(Cl^-)_{outside}$$

Once again, one needs to remember that the above hypothesis has been developed for equilibrium condition. One must also take into consideration that in most metabolically active living cells, no equilibrium will exist between the two sides of the membrane. Metabolism is continuously consuming ions on the inside and, thereby, it is constantly disrupting the equilibrium condition favoring passage of ions from the outside.

3

SOIL SOLUTION

3.1 CONCEPT AND IMPORTANCE OF SOIL SOLUTION

The soil system is composed of three phases: (1) solid, (2) liquid, and (3) gas. The solid phase is a mixture of mineral and organic material and provides the skeletal framework of the soil. Enclosed within this framework is a system of pores, shared jointly by the liquid and gaseous phase. The composition and chemical behavior of the liquid and gas phase are determined by the interaction with the solid phase. The gaseous phase, or soil air, is a mixture of gases. The content and composition of soil air is determined by soil—water—plant relationships. Most biological reactions in soils consume oxygen and produce carbon dioxide, making soil aeration necessary for plant growth. The liquid phase, also designated as the *soil solution*, is composed of water and dissolved substances. The substances are sometimes free salts, and often the ions of the salts are attached to clays, other colloidal material, and/or organic solutes. The water may be free to move depending upon the forces, but the solutes may be more or less constrained or may also effect some constraint on the water.

The soil solution as described above is the medium in which most soil chemical reactions occur. It bathes the plant roots and forms the source from which the roots and other organisms obtain their inorganic nutrients and water. Therefore, the soil solution provides the chemical environment of plant roots, and defining plant—soil-water interrelations in quantitative terms requires a complete and accurate knowledge of soil solution chemistry.

3.2 CHEMISTRY OF SOIL WATER

Water, in its pure state, is a colorless, odorless, and tasteless liquid. It may also exist in a vapor or solid state, and often all three phases can occur at the same time. Some of the properties are peculiar to its liquid state, while other properties may be more common to the vapor and solid states.

Water participates directly in a number of soil and plant reactions, and indirectly affects many others. Its ability to react is determined by its chemical structure. A single water molecule is very small, generally in the dimension of 3 Å (0.3 nm or 3×10^{-8} cm) in diameter. One mole of water (18 ml) contains 6.02×10^{23} individual molecules. An individual water molecule is composed of one atom of oxygen attached to two atoms of hydrogen. The hydrogen atoms are at an angle of about 105° from each other (Figure 3.1). This arrangement causes an imbalance of charges: the center of positive charge being at one end with the center of the negative charge at the other end. Such molecules are called *dipolar* because of their behavior in electrical fields according to those imbalanced charges above. When water crystallizes, the molecules arrange themselves so that hydrogen atoms of one water molecule are located close to an oxygen atom of another molecule of water. The bond by which a hydrogen atom acts as the connecting linkage is called a *hydrogen bond*. As a result of crystallization, water forms a hexagonal structure with many empty spaces. Consequently ice is less dense than liquid water and will float in water. The formation of empty spaces causes a volume expansion of about 9%, and the forces developed by the latter is approximately 150 kg/cm^2. Therefore, when freezing of water occurs in soils and plants, the expansion causes the soil structure to change and the plant cells to rupture.

The presence of dipoles in the water molecules accounts for a number of other important reactions. Cations, e.g., Na^+, K^+, Ca^{2+}, become hydrated through their attraction to the negative pole of water molecules. Likewise, negatively charged clay particles attract water through the positive pole of the water molecule. Polarity of water also encourages dissolution of salts, as the ionic components of salts have

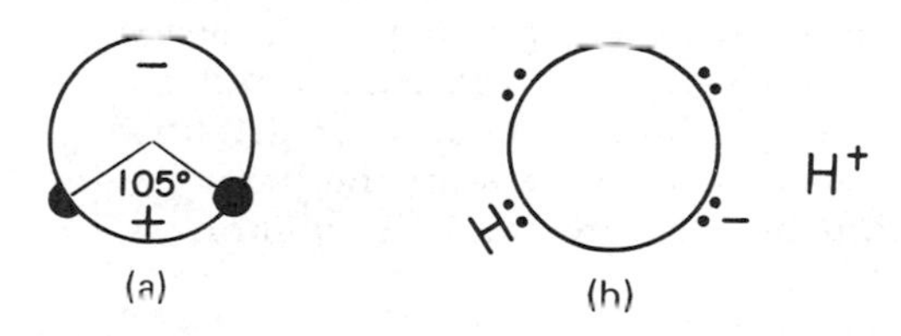

Figure 3.1 (a) A water molecule composed of 2H atoms attached at 105° angle to an O atom; and (b) ionization of water into a hydroxyl (OH^-) and a proton (H^+) or hydrogen ion. Dots are unshared electron pairs, except where H is linked to O.

greater affinity for water molecules than for each other. When water becomes attracted to clay surfaces or ions, they do so in packed clusters. In clusters, the free energy of water is lower than in "free" water, meaning they are less free to move. Free water has greater internal energy than ice or clusters of water. The latter is indicated by the release of 80 cal (334 J) of energy in the form of heat as water freezes into ice. In scientific terms, it means that the entropy of water is lower in the solid than in the liquid state. When ions become hydrated, energy is also released, a phenomenon called *heat of solution*. When clay particles are hydrated, the energy released is called *heat of wetting*. Surface tension is another important property of water that influences its behavior in soils. This property occurs only at liquid-air interfaces, and is the result of greater attraction between water molecules for each other than for the air above. The net effect is an inward force of water molecules at the surface, causing water to behave as if its surface was covered with an elastic membrane. Surface tension plays an important role in soil-water movement called *capillarity*.

3.3 SOIL-WATER ENERGY CONCEPTS

Water in soils may possess different kinds and quantities of energy. Differences in energy content of water at various locations in soils cause it to flow. Retention of water in soils, its uptake and transport in plants, and the loss of water to the atmosphere are all processes which are related to changes in the energy status of soil water. Several kinds of energy are usually involved: potential, kinetic, and electrical energy (Brady, 1974). However, Hillel (1972) indicated that only potential and kinetic energy were traditionally recognized as the two principal forms of energy in physical science. Of the two, potential energy is considered of more importance in determining the state and movement of water in soils. Differences in potential energy from one to another point in the soil tend to create movement of soil water. The water flow is in the direction of decreasing potential energy. The force causing this flow of water is called *water potential difference* (because of a difference between two points). In contrast, a water potential gradient is a continuously changing function of potential in a flow medium. The flow of water continues until the water potential difference between the two points is zero, and an equilibrium condition is attained.

Water Potential (ψ_w)

The term was introduced by Buckingham in 1907, who used it as a synonym for capillary potential (Hillel, 1972). Currently it is applied to describe water's energy status or ability to work. Work is considered performed as water moves from one to another location in soils. At equilibrium condition, water has the potential to do this work, and the energy associated with it is called *potential energy* or briefly *potential.* This water potential is the net effect of several components. It is the sum of the contributions of the various forces acting on soil water, such as matric, osmotic and solute forces. For isothermal conditions the water potential can be formulated as follows:

$$\psi_w = \psi_m + \psi_p + \psi_s \qquad (3.1)$$

where

ψ_m = matric potential
ψ_p = pressure potential
ψ_s = solute potential of soil water

This water potential can assume a positive or a negative value depending on the forces acting on soil water. The presence of solutes and matrix components decreases the capacity of water to perform work; hence under normal field conditions the soil water potential is negative. However, under conditions where the hydrostatic pressure is greater than the atmospheric pressure (in pressure plates), the water potential will be positive.

From the previous discussion, it is apparent that the relative level of energy of water at different locations in soils is of more value to the behavior of soil water than the absolute amount of energy that soil water contains. As indicated earlier, the term water potential is used as a measure to express this energy level of soil water relative to that of water at standard state. If external force fields are excluded, this concept of water potential shows considerable analogy with the concept of chemical potentials. As discussed before, differences in chemical potentials of a component between two points in the soil also govern the direction of movement of the component (in this case the movement of soil water). Not surprising is that a number of authors have tried to apply thermodynamic concepts of chemical potentials in soil—water problems, although several arguments questioning its applicability were frequently reported (Taylor and Ashcroft, 1972; Hillel, 1972; Taylor and Slatyer, 1960). Using thermodynamic principles, the water potential is considered to be equal to the difference between the chemical potential of soil water at an arbitrary equilibrium state and that of soil water at standard state (Taylor and Ashcroft, 1972). That relationship can be expressed as follows:

$$\psi_w = \Delta \mu_w = \mu_w - \mu_w^o \qquad (3.2)$$

where

$\Delta\mu_w$ = difference in chemical potential of soil water

μ_w = chemical potential of soil water at arbitrary equilibrium condition

μ_w° = chemical potential of soil water at standard state

Formerly $\Delta\mu_w$ was called the *moisture potential* (Taylor and Slatyer, 1960).

Total Soil-Water Potential (ψ_t)

This is defined by the International Soil Science Society, Soil Physics Terminology Committee (Aslyng et al., 1963) as the amount of work required to transport reversibly and isothermally an infinitesimal amount of water from a pool of pure water at a specified elevation at atmospheric pressure to the point under consideration. This potential includes the water potential and potentials arising from external force fields. It is generally formulated as follows:

$$\psi_t = \psi_w + \psi_g + \psi_z + \cdots$$

where

ψ_w = water potential

ψ_g = gravitational potential

ψ_z = any potential arising from external force fields

Since in soils, gravity is the only external force field of importance, the total water potential ψ_t is usually expressed as

$$\psi_t = \psi_w + \psi_g \tag{3.3}$$

Depending on the forces acting on soil water, the total water potential can also assume positive or negative values.

Matric Potential (ψ_m)

The attraction of soil solids (matrix) for water provides a matrix force, and that part of the water potential attributed to the matrix force is called *matric potential.* Taylor and Ashcroft (1972) reported that the matric potential is similar to the capillary potential, and replaces terms such as soil moisture tension, soil moisture suction, or matric suction. The matrix force reduces the free energy of the adsorbed water. In the presence of solid particles (matrix), water is subject to adsorption on the particle surfaces, and the adsorbed water cannot move as freely as free water. The matric potential can be determined with a tensiometer.

Pressure Potential (ψ_p)

Pressure differences in soils, resulting from air or pneumatic pressures of the atmosphere on soil water, are the reasons for the development of pressure potentials. In a saturated soil, the pressure potential has a positive value because the hydrostatic pressure is greater than the atmospheric pressure. In a water-unsaturated soil, the pressure potential equals zero, since the liquid (water) pressure can be neglected, while the soil air pressure can be considered equal to the atmosphere. A negative pressure potential occurs only when soil water is subjected to a pressure lower than atmospheric pressure. Taylor and Ashcroft (1972) stated that negative pressures were normally observed in laboratory conditions. However, Hillel (1972) indicated that the negative pressure potential was synonymous to capillary and matric potential, and was of the opinion that both capillary and adsorptive forces were responsible for the negative pressure potential. Thus, the matric potential, or the negative pressure potential, is a reflection of the total effect resulting from the retention of water in soil pores (capillary forces) and on the surfaces of soil particles (adsorption).

Osmotic Potential (ψ_o)

The portion of the water potential attributed to the attraction of solutes (ions and other molecules) for water by osmotic forces is called *osmotic potential*, sometimes also known as *solute potential*. This potential becomes of importance only if a semipermeable membrane is present. The latter acts as a barrier for movement of the solutes, but water can flow freely through the membrane. In the absence of a membrane, the solutes will flow with the water, instead of the solutes attracting water.

Gravitational Potential (ψ_g)

This is the portion of the total water potential attributed to the downward pull of water by gravity. By moving water against the gravitational force, work is performed. The amount of work needed is stored in the water in the form of potential energy. The gravitational potential is independent of the chemical potential but depends on the vertical location and the density of soil water.

Units of Soil-Water Potential

The units for soil-water potential can be expressed in several ways (Table 3.1). Usually the water potential is expressed in units of contained energy per unit mass of water (ergs per gram of water, or joules per kilograms of water). Sometimes, the water potential is stated in units of energy per mole of water (ergs per mole). The latter is also known as *molar water potential.* Another way to express soil water potential is in units of energy per volume of water (ergs per cubic centimeter), which is then called *volumetric water potential.* The latter unit is equivalent to the unit of pressure in physics, dynes per square centimeter (1 erg = 1 dyn/cm).

3.4 PLANT—SOIL-WATER ENERGY RELATION

The theories of water potentials are well established and have proven useful in describing the conditions of water in plant tissue and the movement of water through the plant.

Water is retained in plant cells by adsorptive and osmotic forces. The cells consist of (1) cell walls, which are rigid, but exhibit a capability for elastic expansion; (2) protoplasm, acting as a semipermeable membrane through which water can move freely, and in

Table 3.1 Units of Soil-Water Potentials and Their Equivalents

Energy/unit mass		Volumetric potential, bars	Soil-water suction bars	Relative humidity, %
ergs/g	J/kg			
0	0	0	0	100.00
-1×10^4	-1	-0.01	0.01	100.00
-5×10^4	-5	-0.05	0.05	99.99
-1×10^5	-10	-0.10	0.10	99.99
-3×10^5	-30	-0.30	0.30	99.97
-5×10^5	-50	-0.50	0.50	99.96
-1×10^6	-100	-1.00	1.00	99.92
-5×10^6	-500	-5.00	5.00	99.63
-1×10^7	-1000	-10.00	10.00	

Source: Hillel (1972), Taylor and Ashcroft (1972), and Weast (1972).

contrast with the flow of solutes which are somewhat restricted; and (3) vacuoles, filled with solute-rich cell sap and some colloidal material. The solute and colloidal concentration reduce the activity of water inside the cell. The higher solute and colloid concentration results in attraction of water, and water outside the membrane will move into the cell more rapidly than the solutes can diffuse out. The attraction of the cell for water due to solute concentration in the vacuole is called *solute potential* ψ_s. If the attraction arises from the adsorption of water by colloidal material in the cell, or by protoplasmic colloids, it is called *matric potential* ψ_m. The combination of solute and matric potential is the *osmotic potential* ψ_o. The turgor pressure accompanying the adsorption of water by the cell is called *turgor*, or *pressure potential* ψ_p. This is the potential forcing water out of the cell as a result of inflated conditions of the cell. When water moves into a cell, the cell volume increases (see Figure 3.2) and the protoplast is forced against the cell wall, which, being elastic, expands. The greater the expansion of the cell, the greater will be the pressure exerted on water within the cell, and the turgor pressure increases accordingly. Consequently the flow of water into the cell decreases gradually as the turgor pressure ψ_p within the cell increases. At one point the pressure (turgor) potential can become numerically equal, but opposite in sign, to the combined solute and matric potentials. At the point of equal pressure, the sum of potentials $\psi_p + \psi_s + \psi_m$ is then equal to zero, and the net flux of water into the cell becomes zero. Flux refers to the rate of flow. The sum of the solute, matric, and pressure (turgor) potentials is called the *water potential* ψ_w:

$$\psi_w = \psi_s + \psi_m + \psi_p \tag{3.4}$$

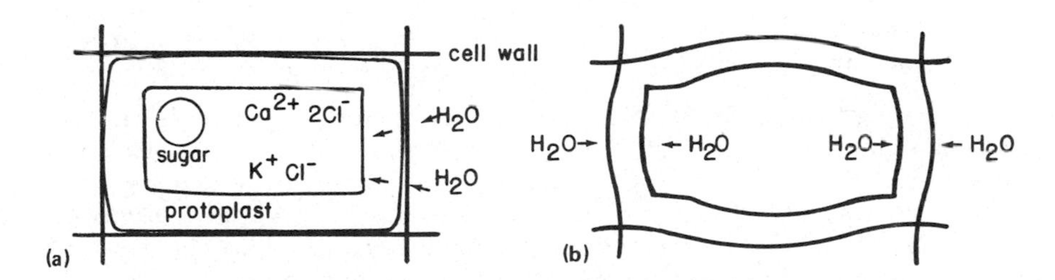

Figure 3.2 (a) Schematic diagram of plant cell showing vacuole filled with cell sap, sugar, Ca^{2+}, K^+, and Cl^- ions. Water moves freely into cell (from Taylor and Ashcroft, 1972); and (b) inflated cell with increased turgor potential exerting pressure on water within the cell. Movement of water into and out the cell stops when turgor equals osmotic potential.

There is also a tendency for water to move through the cell membrane into the soil. This movement can be prevented by increasing the pressure or decreasing the tension on the soil side of the membrane.

3.5 THE LAW OF MASS ACTION AND THE EQUILIBRIUM CONSTANT

Almost all chemical and biochemical reactions occur in aqueous solutions. The processes will be governed by basic chemical laws, one of which is the law of mass action. Consider the reaction

$$A + B \rightleftharpoons C + D$$

in which A and B are the reactants, and C and D are the reaction products. The rate of reaction from left to right (R_1) is proportional to the product of the concentrations of A and B:

$$R_1 = k_1 C_A \times C_B$$

Similarly the rate of reaction from right to left (R_2) can be written as

$$R_2 = k_2\, C_C \times C_D$$

in which k_1 and k_2 are proportionality constants.

At equilibrium R_1 must equal R_2:

$$R_1 = R_2$$

$$k_1\, C_A \times C_B = k_2\, C_C \times C_D$$

$$\frac{C_C \times C_D}{C_A \times C_B} = \frac{k_1}{k_2} = K_{eq} \qquad (3.5)$$

K_{eq} is called the *equilibrium constant*. The above equation is the formulation of the mass action law as first reported by Guldberg and Waage in 1865. A revised definition was, however, provided by van't Hoff in 1877, and since then this law has been applied to various systems. The law of mass action says that at equilibrium the product of the concentrations of the reaction products divided by the product of the concentrations of the reactants is constant in any chemical reaction. In a concentrated solution the concentrations in Eq. (3.5) must be replaced by activities. However, for a first approximation, activities may be replaced by concentrations in moles per liter. The equilibrium constant K_{eq} depends on the temperature of the solution, and not on the pressure of the air or the composition of the system. It is fixed for any given temperature. The larger the numerical value of K_{eq}, the greater the tendency for the reaction to proceed in the direction of the reaction products.

The aforementioned law of equilibrium finds extensive application in many ionic solution phenomena, and the equilibrium constant assumes different kind of expressions, e.g., solubility product constant, dissociation constant or ion product constant of water, ionization constant of electrolytes, cation exchange constant, or ion exchange constants.

3.6 SOLUBILITY PRODUCT

The solubility product is the product of ion concentrations in the saturated solution of a difficultly soluble salt. Consider the dissociation of BA occurring as follows:

$$BA \rightleftharpoons B^+ + A^-$$

Since the activity of pure solids is unity at equilibrium, application of the mass action law gives

$$K_{eq} = K_{sp} = (B^+)(A^-) \tag{3.6}$$

In this type of reaction K_{eq} is called the *solubility product constant* and the symbol K_{sp} is used. The negative log of K_{sp} is pK_{sp}:

$$pK_{sp} = -\log K_{sp}$$

The smaller the pK_{sp}, the more soluble the substance. However, if two solutions, each containing one of the ions of a difficultly soluble salt, are mixed, no precipitation will take place unless the product of the ion concentration in the mixture is greater than the solubility product.

In saturated solution the concentration of B^+ ions is equal to that of the A^- ions. Since the salt is completely ionized, the solubility S of the salt can be represented by the individual ion concentration:

$$S = (B^+) = (A^-)$$

By substituting these in Eq. (3.6), K_{sp} can be rewritten as

$$K_{sp} = (B^+)(B^+) = (B^+)^2$$

or

$$K_{sp} = (A^-)(A^-) = (A^-)^2$$

Therefore:

$$(B^+) = (A^-) = \sqrt{K_{sp}} \quad \text{or} \quad S = \sqrt{K_{sp}} \tag{3.7}$$

Solubility and solubility products are of significance in systems in which a solid is in equilibrium with its solution. Such systems are common systems in the soil. Dissolution of primary minerals, formation and dissolution of clay minerals are a few examples. In the decomposition of kaolinite, formation of gibbsite obeys solubility laws. The dissolution of liming materials and fertilizers are additional examples. The amount of Ca^{2+} released from the liming material is governed by the solubility product of the component ions. Prediction can be made as to the concentration of Ca^{2+} liberated from, and made available to plant growth by the use of the laws of solubility.

3.7 DISSOCIATION OF WATER

Water molecules have a slight tendency to break up (dissociate) into hydroxyl (OH^-) and hydrogen (H^+) ions. The dissociation is so slight that only one in 10^7 molecules of water is dissociated at any one time. However, without this slight dissociation, many of the processes and reactions in water would not be possible:

$$H_2O \rightleftharpoons H^+ + OH^-$$

By applying the mass action law, the following relationship is obtained:

$$K_{eq} = \frac{C_{H^+} \times C_{OH^-}}{C_{H_2O}} = 1.8 \times 10^{-16}$$

K_{eq} is called the *dissociation constant* of water. The concentration C of pure water is 55.5 mol/l. By substituting this for C_{H_2O}

$$C_{H^+} \times C_{OH^-} = 1.8 \times 10^{-16} \times 55.5$$

$$K_w = 1.01 \times 10^{-14} \quad \text{at } 25°C = 298K$$

K_w is called the *ion product* of water and is used for the formulation of pH (see Chapter 8 for discussion).

3.8 DISSOCIATION OF STRONG ELECTROLYTES

Strong electrolytes are dissociated in water completely into their ionic components. NaCl in solid form exists even as Na^+ and Cl^- ions. HCl in water is usually completely dissociated into H^+ and Cl^- ions. It is common practice to present the dissociation of HCl as:

$$HCl \rightleftharpoons H^+ + Cl^-$$

However, the reaction is more accurate as follows:

$$HCl + H_2O \rightleftharpoons H_3O^+ + Cl^-$$

H_3O^+ is called a *hydronium* ion.

Strong bases also dissociate completely into ions.

3.9 DISSOCIATION OF WEAK ELECTROLYTES

Weak acids and bases exhibit only a slight dissociation. Acetic acid in water theoretically dissociates into

$$CH_3COOH \rightleftharpoons H^+ + CH_3COO^-$$

By applying the mass action the extent of dissociation can be written as

$$K_{eq} = \frac{(H^+)\ (CH_3COO^-)}{(CH_3COOH)}$$

In this type of reaction, K_{eq} is called *ionization constant* or K_a. At 25°C the ionization constant is

$$K_a = 1.8 \times 10^{-5}$$

When the concentration of the anion is equal to that of nonionized acid, $K_a = (H^+)$. This is for example attained by mixing 0.1 mol sodium acetate with 0.1 mol acetic acid. Under this condition $(CH_3COO^-) = (CH_3COOH) = 0.1$ mol. Therefore:

$$K_a = \frac{(H^+)\ (CH_3COO^-)}{(CH_3COOH)} = 1.8 \times 10^{-5}$$

$$= \frac{(H^+)\ (0.1)}{(0.1)} = 1.8 \times 10^{-5}$$

$$= (H^+) = 1.8 \times 10^{-5}$$

$$pK_a = pH = 5 - \log 1.8$$

$$= 4.74$$

3.10 THE HENDERSON-HASSELBALCH EQUATION

The discussion above is applied for describing ionic properties of amino acids by Henderson-Hasselbalch. If we consider a weak acid

HA, then according to the above, the ionization constant is

$$K_a = \frac{(H^+)(A^-)}{(HA)}$$

Rearranging the equation above gives

$$(H^+) = K_a \frac{(HA)}{(A^-)}$$

By taking the log, the equation is transformed into

$$\log (H^+) = \log K_a + \log \frac{(HA)}{(A^-)}$$

Multiplication with -1 gives

$$-\log (H^+) = -\log K_a - \log \frac{(HA)}{(A^-)}$$

$$pH = pK_a + \log \frac{(A^-)}{(HA)} \quad (3.8)$$

Equation (3.8) is called the *Henderson-Hasselbalch equation* and is used to predict the behavior of ampholytes, such as amino acids and protein, in solution. All amino acids contain ionizable groups that act as weak acids or bases and give or take protons with change in pH. The ionization of such amphoteric compounds follows the Henderson-Hasselbalch equation, which can be written in the following generalized form:

$$pH = pK_a + \log \frac{\text{unprotonated form (base)}}{\text{protonated form (acid)}} \quad (3.9)$$

When the ratio of the concentration of the unprotonated form to that of the protonated form equals 1, the entire log expression becomes zero. Hence pH = pK_a, as shown earlier for weak acids. Consequently pK_a can be defined (considered synonymous) as the pH when the concentration of unprotonated and protonated species are equal. The pK_a also equals pH when the ionizable group is at its best buffering capacity.

3.11 THE EQUILIBRIUM CONSTANT AND ION PAIRS

Ion pairs are defined as pairs of oppositely charged ions, that behave as a thermodynamic entity (Davies, 1962). Strong electrolytes will often not dissociate completely into their component ions. Due to short range interactions between closely adjacent cations and anions, these ions remain strongly attracted to each other. Therefore, a considerable portion may behave as if they were not ionized at all. Pairing of ions can be illustrated as follows with Ca^{2+} and CO_3^{2-}:

$$CaCO_3^{\circ} \rightleftharpoons Ca^{2+} + CO_3^{2-}$$

in which $CaCO_3^{\circ}$ is the Ca^{2+}- CO_3^{2-} pair. The equilibrium constant for such a reaction is

$$K_{eq} = \frac{(Ca^{2+})\,(CO_3^{2-})}{(CaCO_3^{\circ})}$$

Although K_{eq} is formulated in a similar way to that of weak electrolytes, the dissociation of ion pairs is affected by different forces than those in weak electrolytes. The attraction in ion pairs is caused by coulombic forces, whereas in weak electrolytes covalent bonds are the reasons for a weak dissociation.

Soil cations and anions that have been reported to pair extensively are H^+, K^+, Na^+, Ca^{2+}, OH^-, HCO_3^-, CO_3^{2-}, and SO_4^{2-}. Chlorine, Cl^-, ions do not form ion pairs with other cations to any measurable amounts (Garrels and Christ, 1965, Davies, 1962).

3.12 THE EXCHANGE CONSTANT AND ION EXCHANGE

Negatively charged organic and inorganic soil colloids have the capacity to adsorb and exchange cations, a topic which will be discussed in detail in the chapter of cation exchange. An example of an exchange reaction involving monovalent ions can be illustrated as follows:

$$\boxed{\text{micelle}}\,A + B^+ \rightleftharpoons \boxed{\text{micelle}}\,B + A^+$$

Application of the mass action law gives

$$K_{ex} = \frac{(A^+)\,[B^+]}{[A^+]\,(B^+)}$$

where

( ) = free ion activites
[ ] = adsorbed ion activities
K_{ex} = exchange constant

Rearranging the formula shows the following relation:

$$\frac{(A^+)}{(B^+)} = K_{ex}\frac{[A^+]}{[B^+]}$$

For monovalent cation exchange, the exchange constant K_{ex} is considered a measure of the selectivity of the exchanger or the micelle (clay particle). The *ion selectivity* is defined as the tendency to adsorb one ion more strongly than another. Suppose $K_{ex} = 5$ and the ratio $(A^+)/(B^+)$ of the ion activities in solution equals 1, then one of the ions will be adsorbed more strongly by the micelle than the other. Equal amounts of adsorption of A^+ and B^+ ions take place only if the activity ratio of the ions in solution has the same value as the exchange constant K_{ex}.

3.13 RELATIONSHIP BETWEEN EQUILIBRIUM CONSTANT AND CELL OR ELECTRODE POTENTIAL

If the equilibrium reaction $A + B \rightleftharpoons C + D + ne$ is considered, then in accordance to the concepts of the oxidation potentials the following relationship is valid:

$$E_h = E^\circ + \frac{RT}{nF} \ln \frac{(C)(D)}{(A)(B)}$$

Since

$$\ln \frac{(C)(D)}{(A)(B)} = \ln K_{eq}$$

therefore

$$E_h = E^\circ + \frac{RT}{nF} \ln K_{eq} \qquad (3.10)$$

where

E_h = oxidation potential
E° = standard oxidation potential
R = gas constant
T = absolute temperature
n = valence
F = Faraday constant
K_{eq} = equilibrium constant

Changing from the natural to the common logarithm ($\ln = 2.303 \log$), E_h assumes the following formula:

$$E_h = E^\circ + \frac{0.059}{n} \log K_{eq}$$

3.14 EQUILIBRIUM CONSTANT AND FREE ENERGY RELATIONSHIP

The derivation of the laws governing equilibrium constants comes from thermodynamics. Chemical thermodynamics is the science of energy relationships within chemical systems. In any chemical reaction, energy changes are occurring. A system, which is not in equilibrium, will spontaneously undergo changes by releasing energy. At equilibrium the energy changes of the reactants must equal the energy changes of the products, and the following relationship is valid:

$$\Delta G_r = \Sigma \text{ free energy products} - \Sigma \text{ free energy reactants} = 0 \quad (3.11)$$

Equation (3.11) expresses the first law in thermodynamics; ΔG_r is the free energy change of reaction. The use of the symbol G is preferred by many authors over the symbol F since G denotes the Gibbs free energy.

For a general reaction $A + B \rightleftharpoons C + D$, the free energy change of reaction in thermodynamics is written as

$$\Delta G_r = \Delta G_r^\circ + RT \ln \frac{(C)(D)}{(A)(B)}$$

or

$$\Delta G_r = \Delta G_r^\circ + RT \ln K \quad (3.12)$$

where

K = the activity ratio
ΔG_r° = standard free energy change of reaction

If ΔG_r has a negative value, the reaction will go spontaneously to the right. However, if ΔG_r is positive, the reaction will occur to the reverse direction or to the left. If, on the other hand, $\Delta G_r = 0$, then in accordance to the first law of thermodynamics, the reaction is at equilibrium. At equilibrium the activity ratio equals the equilibrium constant K_{eq}. K_{eq} is also known as the thermodynamic equilibrium constant. Consequently, at equilibrium condition the following relationship is valid:

$$\Delta G_r = 0 = \Delta G_r^\circ + RT \ln K_{eq}$$

therefore

$$\Delta G_r^\circ = -RT \ln K_{eq} \quad (3.13)$$

or

$$\Delta G_r^\circ = -1.364 \log K_{eq} \qquad 25°C = 298 \text{ K}$$

3.15 ACTIVITY AND STANDARD STATE

Activity is a measure of the effective concentration of a reactant or product in a chemical reaction. The concentration of a substance does not always accurately describe its reactivity in a chemical reaction. The activity or effective concentration differs from the actual concentration because of interionic attraction and repulsion. The difference between activity and concentration becomes substantially large when the concentration of the reactants is large. At high concentrations, the individual particles of the reactants may exert a mutual attraction to each other, or exhibit interactions with the solvent in which the reaction takes place. On the other hand, in very dilute condition, interactions are less, if not negligible. In order to correct for the difference between actual and effective concentration, the activity coefficient (γ) is introduced. The activity coefficient expresses the ratio of activity to concentration:

$$\frac{a_A}{c_A} = \gamma \qquad \text{or} \qquad a_A = \gamma c_A$$

where

γ = activity coefficient
a_A = activity of species A
c_A = concentration of substance A

The activity coefficient is not a fixed quantity, but varies in value depending on the conditions. In very dilute (infinite dilution) condition, the activity coefficient approaches unity. The value of $\gamma \sim 1.0$, and hence activity equals concentration:

$$a_A = c_A$$

Activity coefficients apply to cations as well as to anions:

$$\text{Cations:} \quad \gamma_+ = \frac{a^+}{m^+}$$

$$\text{Anions:} \quad \gamma_- = \frac{a^-}{m^-}$$

where

m^+ = concentration of cations
m^- = concentration of anions

The mean ionic activity coefficient is then

$$\gamma_\pm = \left[\left(\frac{a^+}{m^+}\right)\left(\frac{a^-}{m^-}\right)\right]^{\frac{1}{2}} \qquad \text{or} \qquad \gamma_\pm = \left[(\gamma_+)(\gamma_-)\right]^{\frac{1}{2}}$$

Activity coefficient in the standard state is indicated by γ^o. A standard state is defined for each substance in terms of a set of reference conditions. Each pure substance in its standard state is assigned an activity of unity. The standard state of solids and liquids is usually chosen as the pure substance under standard conditions of 1 atm pressure and a specified temperature. Since 298.15 K, equivalent to 25.0°C, is a commonly used temperature, it is called the *reference temperature*. The standard state of gases is a perfect gas, obeying PV = nRT, at 1 atm. pressure and a specified temperature.

3.16 DEBYE-HÜCKEL THEORY AND ACTIVITY COEFFICIENTS

The individual ion activity coefficient can be calculated using the Debye-Hückel equation:

$$-\log \gamma_i = \frac{A z_i^2 \sqrt{I}}{1 + a_i^o B \sqrt{I}} \tag{3.14}$$

where

A, B = constants of the solvents at specified temperature and pressure
z = valence
I = ionic strength
a = effective diameter of the ion

Values of A and B as function of temperature at 1 atm are given below:

Temperature		A	B ($\times 10^{-8}$ cm)
0°C	273°K	0.4883	0.3241
5	278	0.4921	0.3249
10	283	0.4960	0.3258
15	288	0.5000	0.3262
20	293	0.5042	0.3273
25	298	0.5085	0.3281
30	303	0.5130	0.3290
35	308	0.5175	0.3297
40	313	0.5221	0.3305

Source: Manov et al. (1943), Garrels and Christ (1965). Values for a_i^o can be found in Klotz (1950).

When the ionic strength becomes very small, Eq. (3.14) changes into $-\log \gamma_i = A z_i^2 \sqrt{I}$.

3.17 IONIC STRENGTH

The concept of ionic strength was introduced by Lewis and Randall (1921) to assess the combined effect of the activities of several electrolytes in solution on a given electrolyte. It is a useful relation in comparing solutions of diverse composition, as is the case in soil water, in river water, and in lake water. The *ionic strength* is defined as

$$I = \frac{1}{2} \sum m_i z_i^2 \qquad (3.15)$$

where

m = moles of ions
z_i = charge of the ions
I = ionic strength

The summation is taken over all ions, positive and negative. For example the ionic strength of a 1 molar solution of $CaCl_2$ is

$$I = \tfrac{1}{2}[(m_{Ca^{2+}} \times 2^2) + (m_{Cl^-} \times 1^2]$$

$$= \tfrac{1}{2}[(1 \times 4) + (2 \times 1)] = 3$$

That of a $\frac{1}{2}$ molar NaCl solution is

$$I = \tfrac{1}{2}[(m_{Na^+} \times 1^2) + (m_{Cl^-} \times 1^2)]$$

$$= \tfrac{1}{2}[(\tfrac{1}{2} \times 1) + (\tfrac{1}{2} \times 1)] = \tfrac{1}{2}$$

Practice Try to calculate the ionic strength of the following soil solution, for which chemical analysis revealed ion concentrations as listed:

Ion	Parts per million	Moles	$m_i z_i^2$
Na^+	2300	$\frac{2300}{23 \times 10^3} = 0.100$	$0.100 \times 1^2 = 0.100$
Ca^{2+}	80	$\frac{80}{40 \times 10^3} = 0.002$	$0.002 \times 2^2 = 0.008$

Practice (Continued)

Ion	Parts per million	Moles	$m_i z_i^2$
Mg^{2+}	48	$\frac{48}{24 \times 10^3} = 0.002$	$0.002 \times 2^2 = 0.008$
SO_4^{2-}	288	$\frac{288}{96 \times 10^3} = 0.003$	$0.003 \times 2^2 = 0.012$
Cl^-	1750	$\frac{1750}{35 \times 10^3} = 0.050$	$0.050 \times 1^2 = 0.050$
CO_3^{2-}	60	$\frac{60}{60 \times 10^3} = 0.001$	$0.001 \times 2^2 = 0.004$
HCO_3^-	2745	$\frac{2745}{61 \times 10^3} = 0.045$	$0.045 \times 1^2 = 0.045$

$$I = \tfrac{1}{2}(0.100 + 0.008 + 0.008 + 0.012 + 0.050 + 0.004 + 0.045) = 0.114.$$

Table 3.2 Effect of Ionic Strength on Single-Ion Activity Coefficient

	Ionic strength of solution				
a[a]	0.001	0.005	0.01	0.05	0.10
			Monovalent ion activity		
3	0.964	0.925	0.899	0.805	0.755
4	0.964	0.927	0.901	0.815	0.770
9	0.967	0.933	0.914	0.860	0.830
			Divalent ion activity		
5	0.868	0.744	0.670	0.465	0.380
6	0.870	0.749	0.675	0.485	0.405
8	0.872	0.755	0.690	0.520	0.450

Table 3.2 (Continued)

	Ionic strength of solution				
a[a]	0.001	0.005	0.01	0.05	0.10
			Trivalent ion activity		
4	0.725	0.505	0.395	0.160	0.095
9	0.738	0.540	0.445	0.245	0.180

[a]a = effective diameter.

Monovalent, a = 3: K^+, Cl^-, Br^-, I^-, NO_3^-; a = 4: Na^+; a = 9: H^+.

Divalent, a = 5: Sr^{2+}, Ba^{2+}, Ra^{2+}, Cd^{2+}; a = 6; Ca^{2+}, Cu^{2-}, Zn^{2+}, Mn^{2+}, Fe^{2+}; a = 8: Mg^{2+}.

Trivalent, a = 4: PO_4^{3-}; a = 9: Al^{3+}, Fe^{3+}, Cr^{3+}.

Source: Klotz (1950).

The average ionic strength for water in rocks is about 0.100, whereas streams and lakes have ionic strengths of about 0.010. The ionic strength of ocean waters is approximately 1.0 (Garrels and Christ, 1965). The higher the ionic strength, the lower will be the ion activity (lower γ), as illustrated by the data in Table 3.2.

4

COLLOIDAL CHEMISTRY OF SOIL CONSTITUENTS

4.1 THE COLLOIDAL SYSTEM

A *colloid* is a state of matter consisting of very fine particles that approach but never reach molecular sizes. The upper size limit of colloids is 0.2 μm, and the lower size limit is approximately 50 Å (5) nm), the size of a molecule. Plant and soils contain large amounts of solid material in a colloidal state. Such material exhibits chemical and physical properties that depend upon the colloidal condition. The inorganic fraction of soils is made up of boulders, rocks, gravel, sand, silt, and clay. Clay comprises all inorganic solids smaller than 0.002 mm (2 μm) in effective diameter and is considered a colloid. Soil organic matter and plant solids also occur in the colloidal state. Humus, protoplasm, and cell walls exhibit many properties of colloidal systems.

Colloidal systems can be divided into two groups. The colloid is considered *lyophobic* (solute hating) if the dispersed phase does not interact with the dispersion medium. It is called *lyophilic* if it does interact. If the dispersion medium is water, often the terms *hydrophobic* and *hydrophylic* may be used. A hydrophobic colloid can be flocculated, but a hydrophylic usually cannot. Plant colloids are mostly hydrophylic, whereas clay minerals are hydrophobic. The colloidal particle surrounded by its hydration shell is called a *micelle*.

A number of chemical and biological reactions occur at the solid-liquid interfaces. Adsorption takes place at the interface. This refers to the concentration of materials at the colloidal surface. In contrast, the term *absorption* indicates the uptake and retention of one material within another. Sometimes, it is difficult to distinguish between adsorption and absorption, and in such a case the term *sorption* is proposed. Desorption is used to indicate the release or removal of materials that were adsorbed. The substance sorbed is called the *sorbate*, and the material in which sorption occurs is called the *sorbent*.

4.2 THE ORGANIC COMPONENTS

The organic components of soils orginate from the biomass which is characteristic for an active soil. Although, strictly speaking, both living organisms and the dead components are included in soil organic matter, only the nonliving fraction will be discussed here. The non-living organic components are formed by chemical and biological decay of mainly plant materials. They can be divided into (1) materials in which the anatomy of the plant substance is still visible and (2) completely decomposed materials. The first group is of significance in soil physics, e.g., protection of soils by leaf mulch, decreasing bulk density, or soil structure. However, from the standpoint of soil chemistry, the nondecomposed organic fraction is chemically of minor importance because its intact structure exhibits a relatively small surface area, rendering them inactive as adsorbents. Also of major importance are the decomposition products, although their nature and accumulation in soils depend on the types and quantity of plant material subjected to decomposition.

The plant tissue is composed of CHONSP and a number of other elements. The inorganic ions make up the ash content that accounts sometimes for as much as 10% of the dry weight of the tissue. The organic part of the plant tissue is composed of a large number of organic compounds, but only a few are present in detectible amounts in soils after decomposition. They are primarily (1) carbohydrates, (2) amino acids and proteins, (3) lipids, (4) nucleic acids, (5) lignins, and (6) humus.

Carbohydrates

Carbohydrates are by definition polyhydroxy aldehydes, ketones, or substances that yield one of these compounds on hydrolysis. Glucose ($C_6H_{12}O_6$) and fructose ($C_6H_{12}O_6$) are examples of an aldose and a ketose, respectively:

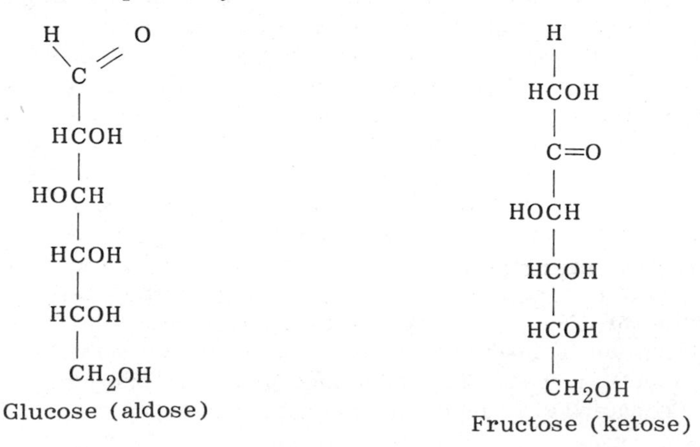

Glucose (aldose) Fructose (ketose)

The term *carbohydrate* indicates that these compounds could be represented by hydrates of carbon: $C_x(H_2O)_y$. However, it was found that this definition was not suitable, since a number of compounds exist with the properties of carbohydrates but does not have the required ratio of hydrogen to oxygen of 2:1. The sugar deoxyribose ($C_5H_{10}O_4$), which is a constituent of deoxyribonucleic acid, a component of every plant cell, is an example. Some of the carbohydrates may also contain N and S and their formula does not agree with $C_x(H_2O)_y$.

Carbohydrates can be divided into three groups: (1) monosaccharides, (2) oligosaccharides, and (3) polysaccharides. Monosaccharides are simple sugars that cannot be hydrolyzed into smaller molecules under reasonably mild conditions. According to the number of carbon atoms, monosaccharides may be trioses ($C_3H_6O_3$), tetroses, and so on up to octoses or nonoses. Oligosaccharides are compound sugars that upon hydrolysis yield two to six molecules of simple sugars. A disaccharide, for example, hydrolyzes into two monosaccharides and upon hydrolysis, pentosaccharides yield five monosaccharides. Polysaccharides are groups of compounds that yield a large number of different monosaccharides upon hydrolysis. They include cellulose and hemicellulose. Some of the monosaccharides that are bonded together by glucosidic bonds to form polysaccharides are glucose, xylose, and arabinose.

The properties of these carbohydrates change significantly with increasing molecular complexicity. The simple sugars are readily soluble in water. The oligosaccharides are crystalline compounds, readily soluble in water, usually having a sweet taste. Polysaccharides are frequently tasteless, insoluble in water, and amorphous in nature. They have high molecular weights. The plant starches and animal glycogens are important examples of polysaccharides. In water they exhibit imbibitions or become dispersed, but they are not strictly soluble. Cellulose, another polysaccharide, is insoluble in water. The molecular weight of cellulose varies from 200,000 to 2,000,000. Polysaccharides are sometimes divided into homo- and heteropolysaccharides. Homopolysaccharides are composed of a repeating monosaccharide, whereas heteropolysaccharides are made up of two (or more) different monosaccharides. The monosaccharide molecules can be bonded in a straight chain, or they can form branchlike structures (see Figure 4.1). Hydrolysis in an acid medium is usually employed to release the monosaccharides. The sugars released are identified either by gas chromatography or by paper chromatography.

Soil polysaccharides may be different than the original plant polysaccharides. They are subject to decomposition by microbial attack, since they are sources of food and energy. Enzymatic attack involves transformation through glycosyl transfer. Two broad types of enzymes have been reported in this respect, e.g., endo- and exoenzymes. The endoenzymes effect the catalytic cleavage of the

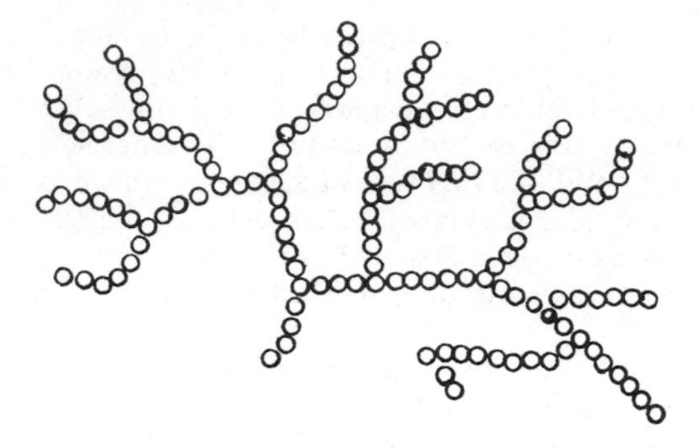

Figure 4.1 Glycogen, a branched structured polysaccharide from animal tissue. Each circle represents a glucose molecule in the chain.

glucosidic bonds, whereas the exoenzymes induce the cleavage of terminal residues. The greater the different types of linkage and the greater the branching of the polysaccharide structure, the greater will be the resistance to enzymatic degradation of soil polysaccharides. This resistance is perhaps the reason why these compounds can accumulate in soils, although the amount rarely accounts for more than 20% in the soil.

Soil polysaccharides can also be protected against degradation by interaction with other soil constituents, such as clay and metal cations. The intimate association with soil clays has been reported to slow down chemical degradation, while adsorption of polysaccharides by, especially expanding, clays [e.g., montmorillonite (Olness and Clapp, 1973, 1975)] in intermicellar spaces renders them inaccessible to enzymatic and/or other microbial attack. Evidence has also been presented that complex reaction with metal cations, e.g., Cu, Fe, and Zn, may inhibit enzymatic decomposition of soil polysaccharides (Martin et al., 1966).

Soil polysaccharides influence soil physical conditions, cation exchange reactions, retention of anions, carbon metabolism, biological activity, and complex reactions of metals. They also react with lignin and amino acids and, therefore, contribute toward the formation of humus, humic acids, and related compounds. Mention has been made in the literature that interaction of soil polysaccharides with soil particles encourages soil aggregation with the consequent formation of granular to crumb structures (Greenland, et al., 1961, 1962; Baver, 1963). Baver (1963) also indicated that the oxidative destruction of soil polysaccharides resulted in a 30 to 90% reduction of stability of soil aggregates. This stabilizing effect on soil structure is attributed to an increase in cementation effect. By interaction with soil clays, the polysaccharide is thought to change the properties of the clay surfaces with respect to adsorption of water. The organic compounds compete with water molecules for adsorption sites and reduce wetting and swelling, thereby increasing cementation. The stabilizing effect of soil aggregates by fungal mycelia, as frequently postulated by a number of authors, is considered a temporary effect by Baver (1963), since mycelia and cells undergo further microbial decomposition.

Amino Acids and Protein

Amino acids are the fundamental structural units of protein. The nitrogen in amino acids occurs as an amino (NH_2) group attached to the C chain. The acid part consists of a terminal C linked to an O atom and an OH group, often written as $-COOH$. The latter, called *carboxyl group*, exhibits acidic properties, since the H in the OH radical is capable of reacting with bases. The general formula of amino acids may be written as

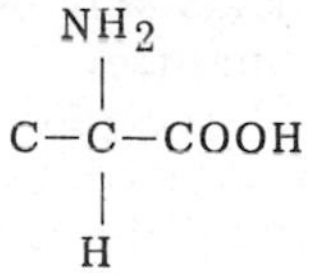

```
     NH2
      |
  C—C—COOH
      |
      H
```

Because the amino group is on the carbon adjacent to the carboxyl group (the α carbon), amino acids with this general formula are called α-amino acids.

The amino acids obtained on hydrolysis of proteins may be classified into (1) aliphatic, (2) aromatic, and (3) heterocyclic amino acids.

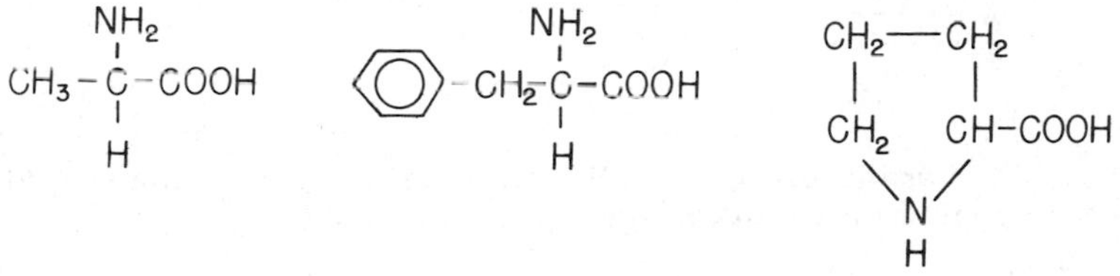

```
      NH2                      NH2            CH2——CH2
       |                        |              |     |
  CH3-C-COOH      (C6H5)-CH2-C-COOH          CH2   CH-COOH
       |                        |                \  /
       H                        H                 N
                                                  H
```

Aliphatic
L-alanine

Aromatic
L-phenylalanine

Heterocyclic
L-proline

Proteins are complex combinations of amino acids. Under refluxing with 6 N HCl for 18 to 24 hr, the protein may be hydrolyzed to its constituent amino acids. Twenty-one amino acids are usually found as protein constituents. The protein is formed by the linkage of many amino acids through the amino and carboxyl groups:

$$\underset{\substack{\text{Glycine}\\ \text{(amino acid)}}}{H_2N{-}\overset{\overset{\large H}{|}}{\underset{\underset{\large H}{|}}{C}}{-}\overset{\overset{\large O}{\|}}{C}{-}OH} + \underset{\substack{\text{Glycine}\\ \text{(amino acid)}}}{H{-}\overset{\overset{\large H}{|}}{N}{-}\overset{\overset{\large H}{|}}{\underset{\underset{\large H}{|}}{C}}{-}\overset{\overset{\large O}{\|}}{C}{-}OH} \rightarrow \underset{\substack{\text{Peptide}\\ \text{(protein)}}}{H_2N{-}\overset{\overset{\large H}{|}}{\underset{\underset{\large H}{|}}{C}}{-}\overset{\overset{\large O}{\|}}{C}{-}N{-}\overset{\overset{\large H}{|}}{\underset{\underset{\large H}{|}}{C}}{-}\overset{\overset{\large O}{\|}}{C}{-}OH} + H_2O$$

Since the N content of most proteins is about 16% and since this element is easily analyzed as NH_3 by the Kjeldahl procedure, the protein content can be estimated by determination of the N content and multiplying by 6.25 (100/16).

Zwitterion

The amino acids, with certain exceptions, are generally soluble in water and are insoluble in nonpolar organic solvents such as ether, chloroform, and acetone. Since amino acids contain both a carboxyl and an amino group, these compounds will react with acids and alkalis. Such compounds are said to be amphoteric by nature. If, for example, alanine is dissolved in H_2O, the pH is 7.0. If electrodes are placed in the solution and a potential difference is placed across the electrodes, the amino acid will not migrate in the electric field. However, if alkali is added to the solution, alanine becomes negatively charged and will migrate to the positive anode. Similarly, when acid is added, alanine becomes positively charged and migrates to the negative cathode. This behavior can be explained by considering alanine a zwitterion (Bjerrum, 1923):

$$H_3C{-}\overset{\overset{\large NH_3^+}{|}}{\underset{\underset{\large H}{|}}{C}}{-}COO^-$$

At pH 7.0, the amino group is still protonated. When alkali is added, the excess proton on the amino group is neutralized ($pK_a = 9.7$):

$$H{-}\overset{\overset{\large COO^-}{|}}{\underset{\underset{\large H}{|}}{C}}{-}NH_3^+ + OH^- \rightarrow H_2N{-}\overset{\overset{\large COO^-}{|}}{\underset{\underset{\large CH_3}{|}}{C}}{-}H + H_2O$$

When acid is added, the dissociated carboxyl group accepts the proton:

$$H_3C{-}\overset{\overset{\large NH_3^+}{|}}{C}{-}COO^- + H^+ \rightarrow H_3C{-}\overset{\overset{\large NH_3^+}{|}}{C}{-}COOH$$

At pH = 2.3, the carboxyl group is half protonated.

Lipids

Lipids are heterogeneous compounds of fatty acids, waxes, and oils. The term *lipid* does not imply a particular chemical structure. The name is used to describe substances that are soluble in fat solvents, such as ether, chloroform, or benzene.

The lipids can be classified into (1) neutral lipids (glycerol), (2) phosphatides, (3) glycolipids, and (4) terpenoid lipids, including carotenoids and steriods. They have limited solubility in water and exhibit a hydrophobic character. The P-containing lipids are also called *phospholipids*. Many of the lipids in plant and animals are associated with proteins and carbohydrates.

Nucleic acids

Each plant and animal cell contains a discrete rounded or spherical body, called the nucleus, which contains nucleic acids. Nucleic acids, first isolated in 1969 by F. Miescher, are polymers with high molecular weights. Their repeating unit is a mononucleotide rather than an amino acid. These acids control the synthesis of enzymes and proteins and are also responsible for the genetic transfer in cell division. Two types of nucleic acids are generally recognized: (1) deoxyribonucleic acid (DNA), a constituent of cell nuclei, and (2) ribonucleic acid (RNA), located in the nucleolus and in the cytoplasmic nuclear membrane, called endoplasmic reticulum. Both DNA and RNA consist of long chains of alternating sugar and phosphate residues. In RNA, the sugar is d-ribose. The sugar in DNA, as the name implies, is 2-deoxyribose.

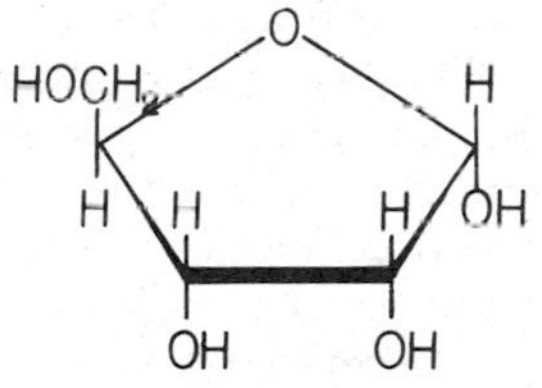

α-D-ribose α-2-deoxyribose

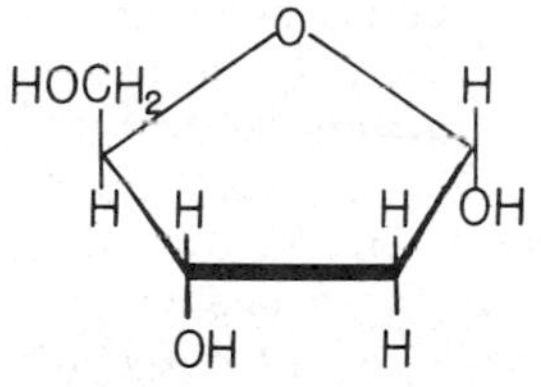

In most cells, these nucleic acids are conjugated with proteins to form nucleoproteins. Nucleoprotein containing DNA, a major component of chromosomes, determines genetic heredity. On the other hand, nucleoproteins containing RNA, known as ribosomes, are important in protein synthesis. Three groups of ribosomal RNA are identified: (1) rRNA (ribosomal RNA), which is the predominant group, amounting frequently to 80% of the total RNA content (2) soluble RNA (sRNA), or sometimes also called transfer RNA (tRNA), and (3) messenger RNA (mRNA), which usually occurs in low concentration. Soluble RNA carries amino acids to their specific sites on the protein template and is, therefore, considered an amino acid carrier or amino acid adaptor, whereas mRNA acts as the messenger of DNA. During the formation of protein, mRNA directs the linkage of amino acids with tRNA.

Lignins

Lignin is a system of thermoplastic tridimensional polymers derived from coniferyl alcohol or guaiacyl propane monomers. Plant lignins can be divided into three types: (1) lignin from soft wood, (2) lignin from hard wood, and (3) lignin from grasses, bamboo, and palm (Figure 4.2). These combinations form large, complex molecules. Many of the C atoms are connected to OH radicals (phenolic hydroxyl groups) in which the behavior of the H is much the same as that in the carboxyl groups of organic acids. The bulk of lignin occurs in the secondary cell wall where it is associated with cellulose and hemicellulose in stems. The quantity of lignin increases with plant age and stem content. It is a very important constituent of woody tissue, and it contains the major portion of the methoxyl content of the wood. It is insoluble in water, in most organic solvents, and in strong sulfuric acid. Lignin has a characteristic UV absorption spectrum and gives characteristic color reactions with many phenols and aromatic amines. It hydrolyzes into simple products, as do the complex carbohydrates and proteins. When oxidized with alkaline benzene, it yields up to 25% vanillin.

Humus and Humic Acids

Terminology and Definitions Soil organic matter is often divided into nonhumified and humified materials. The nonhumified substances are the compounds in plant and other organisms with definite characteristics discussed in the preceding pages, e.g., carbohydrates, amino acids, protein, lipids, nucleic acids, and lignins. These compounds are usually subject to degradation and decomposition reactions. But sometimes they can be adsorbed by inorganic soil components, such as clay, or they may occur in anaerobic conditions. Under such conditions, the compounds above will be relatively protected against decomposition. The humified fraction is known as humus, or currently as humic compounds, and are considered the end product of decomposition of plant material in soils. The term *humic acid* originated with Berzelius in 1830, who classified the soil humic fraction into (1) humic acid, the fraction soluble in bases, (2) crenic and apocrenic acid, the fraction soluble in water, and (3) humin, the insoluble and inert part. Humic acid was also referred to as ulmic acid, whereas humin was also called ulmin by Mulder in 1840. In 1912, Oden proposed the use of the name fulvic acid replacing the terms crenic and apocrenic acids.

Today humic compounds are defined as amorphous, colloidal polydispersed substances with yellow to brown-black color and relatively high molecular weights. A number of authors believe that

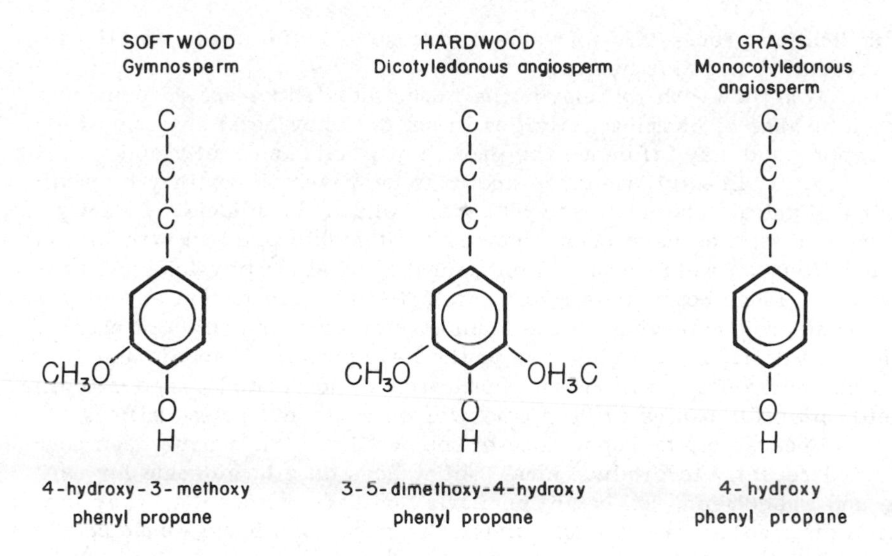

Figure 4.2 Chemical structure of building constituents of lignin from softwood, hardwood, and grass.

these compounds are heterogeneous in molecular weight, although chemically they may be homogeneous in composition (Felbeck, 1965). Based on solubility in acids and alkalis, the humic compounds can be separated into several humic fractions (Flaig et al., 1975):

Fractions	Alkali	Acid	Alcohol
Fulvic acid	Soluble	Soluble	—
Humic acid	Soluble	Insoluble	Insoluble
Hymatomelanic acid	Soluble	Insoluble	Soluble
Humin	Insoluble	Insoluble	Insoluble

According to the German workers, humic acid can be further separated with neutral salt solutions into brown humic acid (soluble in NaCl) and gray humic acid (insoluble in NaCl). In addition to the major fractions above, several authors have also reported the isolation of a green humic acid fraction (Kumada and Hurst, 1967).

Humus or humic substances are very important soil constituents. Depending on climatic conditions and cultural practices, the humus content of soils often stabilizes at a fairly definite amount. In the Southern region of the United States, with a prevailing subtropical climate, soil humus content seldom exceeds 3.5% (Tan et al., 1975), while the carbon to nitrogen (C/N) ratio usually narrows down in a humification process from a value in excess of 20 for fresh material to a value of 8 to 20 for humus.

Together with soil clays, the humic substances are responsible for a number of chemical activities in soils. They enter in complex reactions and may influence the growth of plants indirectly and directly. Indirectly, they are known to improve soil fertility by modifying physical, chemical, and biological conditions in soils. Directly, humic substances have been reported to stimulate plant growth through their effect on metabolism and on a number of other physiological processes. Humic compounds also participate in soil formation and play an important role especially in the translocation or mobilization of clays, aluminum, and iron, giving rise to the development of spodic and argillic horizons. In industry, humic acids and related compounds find application for use as drilling muds for oil wells and as emulsifiers.

Because of the importance in soil fertility, trials have been conducted recently to produce humic substances on a large scale for use as soil amendment, soil conditioner, or fertilizer. Recently, they have been distributed commercially under the names of "liquid humic acid" and "clod buster."

Extraction and Isolation of Humic Substances A number of methods are available for the extraction and isolation of humic substances from soils. The selection of a suitable extractant is based on two conditions: (1) The reagent should have no effect on changing the physical and chemical nature of the substances extracted, and (2) the reagent should be able to quantitatively remove the humic substances from soils. Over the years many inorganic and organic solvents have been evaluated for their effectiveness in extracting humic compounds (Stevenson, 1965; Schnitzer, et al., 1959), usually with mixed results in meeting the two conditions above. Some of the reagents, e.g., dilute bases, can meet the condition set for quantitative removal of humic fractions. However, all of them will have some influence on modifying the physical and/or chemical properties of the extracted substances (Flaig et al., 1975), and the possibility of creating artifacts still confronts the investigator. Some of the reported inorganic reagents used in extraction are

Acids	Bases and salts
0.1 $\underline{N}$ HCl	0.1 $\underline{N}$ $NaOH$
0.025 $\underline{N}$ HF	0.5 $\underline{N}$ $NaOH$
1% H_3BO_3	0.1 $\underline{M}$ Na_2CO_3
	0.1 M NaF
	0.1 M $Na_4P_2O_7$ pH 7.0
	0.1 M $Na_4P_2O_7$ pH 9 to 10
	0.2 M Na_2-EDTA
	0.1 M $Na_2B_4O_7$

Among these reagents, NaOH and $Na_4P_2O_7$ are most widely employed in extraction. Introduced for the first time in 1919 in a generally accepted procedure by Oden, NaOH appears to be the most effective in quantitative removal of humic substances in soils. Since the use of this reagent may induce autooxidation of humic acids, it is usually suggested to conduct extraction with NaOH under a N_2 gas atmosphere. A solution of 0.1 N NaOH is preferred because it is of a milder nature for extraction than 0.5 N NaOH (Pierce and Felbeck, 1975).

Although not as effective as NaOH, $Na_4P_2O_7$ is used frequently for extraction of humic fractions from soils high in sesquioxides content (Kononova, 1961). To increase effectiveness of extraction, a solution with pH 9 to 10 is recommended. Although occasional reports to the contrary have been noted, the use of 0.1 M $Na_4P_2O_7$ often eliminates the need of decalcifying the soil samples prior to extraction as sometimes required with NaOH. In some instances, it has been reported that humic fractions isolated with $Na_4P_2O_7$ exhibited infrared spectra with better resolution than those obtained by NaOH extraction (Tan, 1978c). However, a comparative study on the effectiveness of NaOH and $Na_4P_2O_7$ extraction of humic acids by Orioli and Curvetto (1980) yielded indications that with the pyrophosphate method three high molecular weight fractions of humic acids were not extracted.

Extraction with acids as proposed by Schnitzer et al. (1959) technically yields only fulvic acids, since by definition only fulvic acids are soluble in acids.

The organic solvents used for extraction of humic substances were oxalic acid, formic acid, phenol, benzene, chloroform, or mixtures of these, acetylacetone, hexamethylenetetramine, dodecylsulfate, and urea (Schnitzer and Khan, 1972). Thus far, none of these has been satisfactory. Using 0.5 M and 0.1 M hydroxymethylamine, Orioli and Curvetto (1980) obtained humic acids with different carboxyl contents than those extracted with NaOH. However, no differences were noted in the electrophoretograms of both the humic acids.

The most common procedure for extraction and fractionation of humic acids with NaOH is shown in Figure 4.3.

Chemical Characterization and Composition

Elemental Composition. An example of analytical data of humic and fulvic acids is shown in Table 41. Humic acid is usually rich in carbon, which ranges from 41 to 57%. The lower ranges are exhibited by fulvic acids and humic acids in tropical soils. Fulvic acid distinguishes itself from humic acid also by a higher oxygen, and by lower hydrogen and nitrogen contents. The O content was 44 to 54% of fulvic acids versus 33 to 46% in humic acids. The nitrogen content in fulvic acid shows a range of 0.7 to 2.6% in contrast to humic acids, which contain 2 to 5% N. In the case of humic acids, Flaig (1975) reported that the brown humic acid fraction of Chernozemic soils

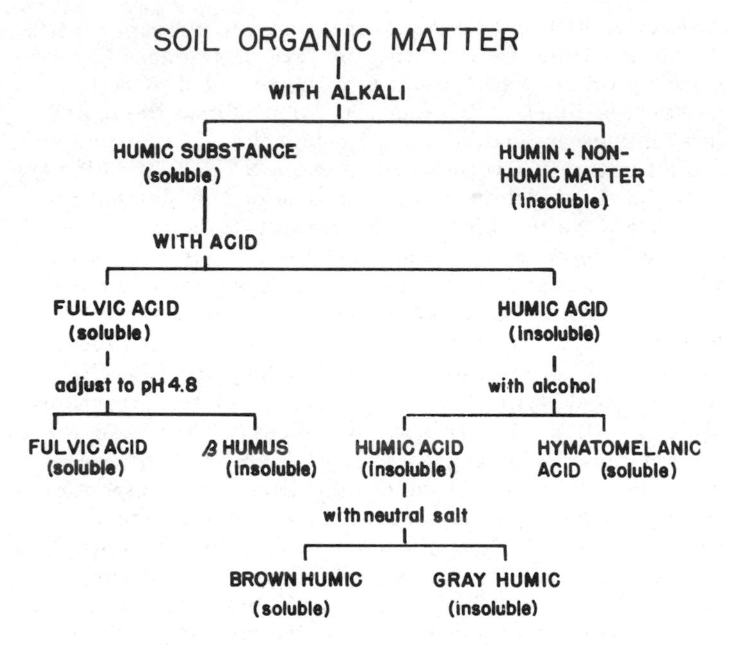

Figure 4.3 Flowsheet for the separation of humic compounds into the different humic fractions.

(Mollisols) was usually higher in nitrogen content than was the gray humic fraction. By mild degradation of humic compounds with peracetic acid oxidation, Schnitzer and Hindle (1980) succeeded in differentiating the N-containing components in the humic molecule. Different types of N components were detected, e.g., NH_4^+-N, NO_2^- + NO_3^--N, amino acid-N, amino sugar-N, and by difference from total N, "unknown" N. These authors reported that 16.6 to 59.1% of the unknown N could be converted into NH_3 and N gases.

The amount of N detected in humic acids indicates the necessary participation of N-containing compounds, e.g., amino acids, amino sugar, in the formation of humus. Electron spin resonance (ESR) analysis reveals that humic acids have an aromatic core, containing physically or chemically bonded proteins, polysaccharides, simple phenols, and metals (Cranwell and Haworth, 1975), and according to Gosh and Schnitzer (1980a), semiquiones also. The linkage to the core renders considerable stability to amino acids, protein, and polysaccharide toward biochemical attack.

Based on the data above, Kononova (1961) reported that the humus nitrogen content in Chernozems (Mollisols) amounted to 7 to 10 t/ha in the top 20 cm. This value decreased toward the Spodosols and in Aridisols, where 2 to 3 t/ha of humus N was reported in the top 20 cm of sierozems. These figures suggest that humic acid may contribute significantly toward the N supply of soils for plant growth.

Total Acidity. The total acidity or exchange capacity of soil humic compounds is attributed to the presence of dissociable protons or hydrogen ions in aromatic and aliphatic carboxyl and phenolic hydroxyl groups. Humic acids are characterized by a lower total acidity and lower carboxyl content than fulvic acids (Table 4.1). The total acidity of humic acids amounts generally to 5 to 6 mEq/g,* with the exception being in humic acids of Spodosols and Ultisols. Spodosols are very acid soils containing humic acids which in some cases exhibit a relatively high total acidity and high carboxyl content. In contrast to humic acids, fulvic acids have a total acidity of 10 to 12 mEq/g which is approximately two times that of humic acids. The carboxyl content of fulvic acid is two to three times higher than that of humic acids, but the phenolic hydroxyl group concentration does not seem to differ significantly from that of humic acids. It can, therefore, be concluded that the higher acidity of fulvic acid is perhaps attributed to the higher carboxyl content. With the exception noted for one of the Spodosols, the data in Table 4.1 also indicate that the variation observed in total acidity of humic compounds is not affected by the variation in soils, but rather by type of organic acid.

The value for total acidity is, however, also dependent on the method of analysis. It is measured usually by the $Ba(OH)_2$ adsorption method, which yields higher values than other methods. As can be noticed from Table 4.2, total acidity measurements using KOH, NaOH, or Ba acetate procedures generally yield lower values (Felbeck, 1965).

Another factor affecting total acidity or exchange capacity of humic substances is the pH. Using titration procedures, Tan (1978a) obtained in his chelation studies exchange capacities of humic acids ranging from 2.26 to 4.48 mEq/g. As can be seen from the data in Table 4.3, the total acidity of humic acid was 2.26 to 2.83 mEq/g at pH 7.0, but increased to 4.43 to 4.48 mEq/g at pH 11.5. These values were in agreement with those reported by Tiurin and Kononova (1962) and Orlov and Erosiceva (1967). The recovery values for Al^{3+}, Fe^{3+}, and Cu^{2+} of 4.4 to 4.9 mEq/g HA (pH 11.5) compared favorably with acidity values of humic acids found by Butler [J. H. A. Butler, unpublished Ph.D. thesis, University of Illinois, 1966; see also Stevenson (1976a)].

Carboxyl Groups. Humic acid is generally characterized by a lower carboxyl group content than fulvic acid. The data in

---

**Note:* The unit used by the International System (SI) to express mEq per 100 g is cmol (p^+) kg^{-1}.

Table 4.1 Chemical Composition of Fulvic (FA) and Humic Acids (HA) Extracted from Tropical and Temperate Region Soils.

	C,%	H,%	O,%	N,%	S,%	Total acidity, mEq/g	Carboxyl, mEq/g	Phenolic hydroxyl, mEq/g	Alcoholic hydroxyl, mEq/g	Carbonyl, mEq/g
HA-Alfisols (temp.)	56.8	5.3	33.3	4.6	n.d.[a]	6.8	3.9	2.8	n.d.	n.d.
HA-Alfisols (trop.)	52.3	5.2	37.2	3.6	n.d.	n.d.	n.d.	n.d.	n.d.	n.d.
HA-Inceptisol (temp.)	51.4	5.8	38.7	4.1	n.d.	6.0	2.4	3.6	n.d.	n.d.
HA-Oxisols (trop.)	44.3	7.7	38.0	2.1	n.d.	n.d.	n.d.	n.d.	n.d.	n.d.
HA-Spodosols (temp.)	56.7	5.2	35.4	2.4	0.4	5.7	1.5	4.2	2.7	0.9
HA-Spodosols (temp.)	49.0	4.6	45.7	0.7	n.d.	12.0	9.2	2.8	n.d.	n.d.
HA-Ultisols (temp.)	48.7	4.8	42.7	3.8	n.d.	8.7	2.7	6.0	n.d.	n.d.
HA-Ultisols (trop.)	44.8	6.3	36.7	2.8	n.d.	n.d.	n.d.	n.d.	n.d.	n.d.
FA-Inceptisol (temp.)	47.9	5.2	44.3	2.6	n.d.	n.d.	n.d.	n.d.	n.d.	n.d.
FA-Spodosols (temp.)	50.9	3.3	44.7	0.7	0.3	12.4	9.1	3.3	3.6	3.1
FA-Spodosols (temp.)	50.2	4.6	43.4	1.8	n.d.	12.1	7.9	4.2	n.d.	n.d.
FA-Ultisols (temp.)	40.6	4.1	53.9	1.4	n.d.	10.2	8.8	1.4	n.d.	n.d.

[a] n.d. = not determined.

Source: Schnitzer and Khan (1972), Schnitzer (1975), Cranwell and Haworth (1975), Martin et al. (1977), and Tan and Van Schuylenborgh (1959).

Table 4.2 Total Acidity Measured by Different Methods

	Total acidity, mEq/g, by			
	$Ba(OH)_2$	KOH	NaOH	Ba acetate
Peat	3.8	2.96	2.54	1.83
Pine Forest soil	6.9	—	—	—
Dunkirk soil (Ontario)	4.3	—	—	—

Source: Felbeck (1953).

Table 4.1 show a carboxyl concentration in humic acids ranging from 1.5 to 2.7 mEq/g HA in contrast with fulvic acid, with a range of 7.9 to 9.1 mEq/g FA. The exception is the carboxyl content of humic acid in one of the Spodosols where a value of 9.2 mEq/g HA was detected. As reported earlier, Spodosols are very acid soils containing acid humus.

A number of methods are available for determination of the amounts of carboxyl groups in humic and fulvic acids, e.g., methods using ion exchange, decarboxylation, iodometry, esterification, and the Ca acetate procedure. Many scientists are using the esterification and Ca acetate methods. Depending on the procedures used, different values are obtained. With esterification, employing diazomethane and dimethylsulfate (Tan, 1975; Schnitzer, 1974; Stevenson and Goh, 1972) to esterify the COOH groups, Felbeck (1965) noted that 54% of the exchange capacity of humic acids was attributed to COOH groups.

Table 4.3 Effect of pH on Amounts of Metal Ions (mEq per 100 mg) Chelated by Humic Acids

		pH	
Titration product	Metal ion recovery, mEq per 100 mg	Before titration	After titration
HA-Al^{3+}	0.445	11.5	7.05
HA-Al^{3+}	0.226	7.0	5.12
HA-Fe^{3+}	0.443	11.5	6.42
HA-Fe^{3+}	0.255	7.0	4.38
HA-Cu^{2+}	0.448	11.5	6.82
HA-Cu^{2+}	0.283	7.0	6.03

Source: Tan (1978a).

The Ca acetate method makes use of the reaction in which acetic acid is formed and released according to the reaction:

$$2R{-}COOH + Ca(CH_3COO)_2 \rightarrow (RCOO)_2Ca + 2CH_3COOH$$

The COOH content is then determined by titration of the acetic acid with a standard base. Using this procedure, Schnitzer and Khan (1972) reported a carboxyl content of 1.5 mEq/g of humic acid. As indicated earlier, fulvic acid distinguishes itself from humic acid by containing a higher carboxyl content amounting to 9.1 mEq/g of fulvic acid.

Hydroxyl Groups. Humic substances contain a variety of hydroxyl groups, but for characterization of humic acids generally three types of OH groups are distinguished:

1. Total hydroxyls, measured by acetylation, are the OH groups associated with all functional groups, such as phenols, alcohols, enols, and hydroquinones. However, in many cases total hydroxyls refer only to the sum of phenolic and alcoholic-OH groups.
2. Phenolic-OH groups are OH attached to phenol groups. The amount can be calculated by difference as follows:

   mEq total acidity - mEq COOH = mEq phenolic-OH

3. Alcoholic-OH groups are OH associated with alcoholic groups. The amount can also be determined by difference as follows:

   mEq alcoholic-OH = mEq total OH - mEq phenolic-OH

The amount of phenolic and alcoholic hydroxyl groups does not differ significantly between humic and fulvic acids (Table 4.1), while the reactivity of alcoholic OH groups is usually considered lower than that of phenolic groups.

In addition to the total acidity, carboxylic and hydroxyl groups, as discussed above, humic compounds also contain carbonyl (C=O) groups, either as ketonic C=O or as quinoid C=O groups. Each of these groups only amounts to approximately 1 mEq/g of humic or fulvic acid (Schnitzer, 1975). In conclusion it can perhaps be stated that fulvic acid is in general more acidic and possesses more carboxyl groups than humic acids. Recently Sposito and Holtzclaw (1977) reported that fulvic acid extracted from sewage sludge was a heterogeneous polynuclear polyacid. They contain functional groups which can behave very strongly or weakly acidic. The strongly acidic groups ionize at $pH < 2.0$ and are perhaps attributed to sulfonic acid groups. The more weakly acidic groups ionize according to the authors above at $pH > 10.0$ and are caused by carboxyls, phenolic-OH, and SH- and N-containing groups. This polyacid nature is considered the reason fulvic acids form weak ion pairs with protons.

Spectral Characteristics of Humic Compounds

Ultraviolet and Visible Light Spectrophotometry. The color of humic substances is a physical property that has attracted the attention of many scientists who attempted to use it for characterization of humic acid fractions in soils (Flaig et al., 1975; Tan and Giddens, 1972; Schnitzer, 1971; Kononova, 1966; Tan and Van Schuylenborgh, 1961). In Germany, especially, color properties of humic substances have attracted a number of investigators. They reported that the rate of light absorption was characteristic for the type and/or molecular weight of humic substances. In UV-visible light spectrophotometry, humic solutions are scanned and the absorbance recorded at various wavelengths between 300 and 800 nm. By plotting the logarithm of the absorbance against the wavelengths, a straight line regression is obtained. The slope of such a line has been used for differentiation of humic substances. Fulvic acids are considered to yield spectra with a steep slope in contrast to humic acids (see Figure 4.4). The slope of the spectral curve can be expressed as a ratio or quotient of the absorbance at two arbitrarily selected wavelengths, e.g., absorbance at 400 and 600 nm, called the *color ratio:*

$$\text{Color ratio:} \quad E_4/E_6 \text{ or } Q_{4/6} = \frac{\text{extinction (absorbance) at 400 nm}}{\text{extinction (absorbance) at 600 nm}}$$

This color ratio is then used as an index for the rate of light absorption in the visible range. A high color ratio, 7 to 8 or higher,

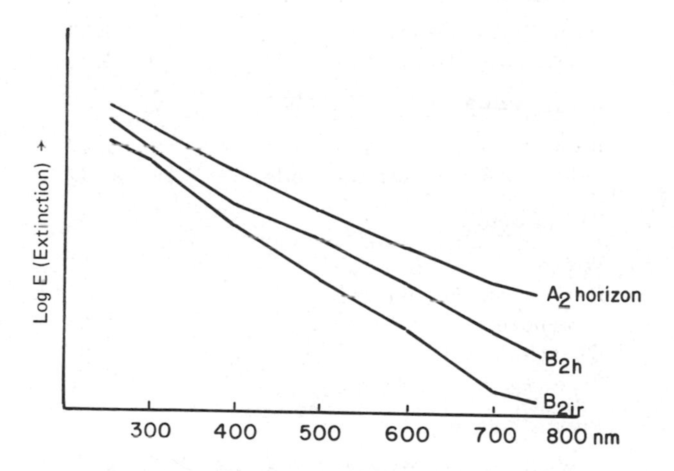

Figure 4.4 Light absorption of humic acids of A and B horizons of a Spodosol in tropical regions. (From Tan and Van Schuylenborgh 1961).

corresponds with curves with steep slopes and is usually observed for fulvic acids or humic fractions with relatively low molecular weights. On the other hand a low color ratio, 3 to 5, corresponds to curves that are less steep in slopes. These curves are noted for humic acids and other related compounds with high molecular weights. The data in Table 4.4 show some E_4/E_6 ratios of humic substances extracted from temperate region soils. It can be noticed that humic acids with high molecular weights (MW > 30,000) have lower E_4/E_6 values (4.32 to 4.45) than humic acids with lower molecular weights (MW = 15,000). The lower molecular weight fractions exhibit E_4/E_6 values of 5.47 to 5.49.

Since UV-visible light spectra of humic compounds are generally featureless lines, Salfeld (1975) proposed the use of differences between two adjacent absorbances. By plotting the logarithms of these differences (ΔE) against the wavelength, a curve was obtained with a number of peaks (see Figure 4.5), called the *derivative spectrum*.

Table 4.4 Color Ratio E_4/E_6 of Humic Substances Isolated from Soils in Temperate Region

Soil taxonomy orders	Great group	E_4/E_6
	Humic acids, MW > 30,000[a]	
Ultisols	Hapludults (Cecil soil)	4.32
Ultisols	Paleudults (Greenville soil)	4.45
	Humic acids, MW = 15,000[a]	
Ultisols	Hapludults (Cecil soil)	5.49
Ultisols	Paleudults (Greenville soils)	5.47
	Humic acids[b]	
Spodosols	Podzol	5.0
Alfisols	Dark Gray Forest soil	3.5
Mollisols	Chernozem	3.3
Mollisols	Chestnut soil	3.9
Aridisols	Sierozem	4.3
	Fulvic acids	
Ultisols[a]	Hapludults (Cecil soil)	8.0
Unknown (Canada)[b]	—	6.0-8.0

[a]From Tan and Giddens (1973).
[b]Adapted from Schnitzer and Khan (1972), and Kononova (1966).

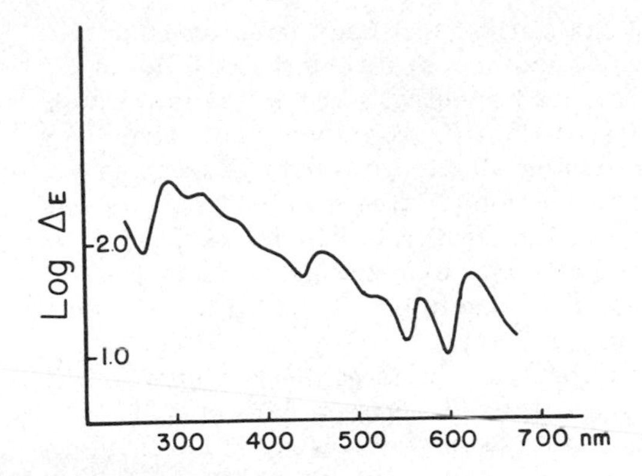

Figure 4.5 Derivative spectrum of humic acid from a Spodosol. [From Salfeld (1975).]

Although the absorption spectra of humic acids in the visible range have the form of straight lines, the inclination or angle of the curves, as expressed by $\Delta \log K = \log K_{400} - \log K_{600}$, has been used in Japan for characterization and distinction of several types of humic acids (Kumada, 1965, Kumada and Miyara, 1973; Yoshida et al., 1978). Four types of humic acid have been recognized:

	Δlog K	Occurrence
A-type humic acid	<0.6	A horizons of volcanic ash soils
B-type humic acid	0.6 to 0.8	Brown forest soils, red-yellow soils and paddy soils
R_p-type humic acid	0.8 to 1.1	Peat, decomposed grasses, stable manure
P-type humic acid	Absorption bands near 615, 570, and 540 mμm (nm) occur mainly in Podzols (Spodosols). This type of humic acid can be distinguished again in Pg (green) and P_b (reddish-brown) humic acid fractions.	

Finally, it should also be mentioned that humic compounds may exhibit fluorescence spectra. Using fluorescence excitation spectroscopy, Gosh and Schnitzer (1980b) indicated that both fulvic and humic acids showed curves with distinct bands at 465 nm or mμm. However, fulvic acid appeared to distinguish itself from humic acid by displaying an additional band at 360 nm.

Infrared Spectroscopy. This method has been used extensively in the past to characterize humic substances, although some doubt exists as to the usefulness of infrared spectra. The latter is perhaps caused in part by the complexity of the infrared spectra of humic preparations and by questions arising about the purity of samples (Tan, 1976b). In spite of these questions, infrared spectroscopy has proven to be very useful and can identify three different types of humic fractions, e.g., fulvic, humic, and hymatomelanic acids [see Figure 4.6 and Schnitzer (1971), Stevenson and Goh (1972), Stevenson and Goh (1971), Goh and Stevenson (1971)].

As can be noticed from the spectra, fulvic acid has a strong absorption band at 3400 cm^{-1}, a weak band between 2980 and 2920 cm^{-1}, and a medium strong band at 1720 cm^{-1} followed by a shoulder at 1650 cm^{-1} attributed to vibrations of OH, aliphatic C—H, carbonyl (C=O), and carboxyl groups in COO^- form, respectively. The strong band at 1000 cm^{-1} is not necessarily caused by contamination with

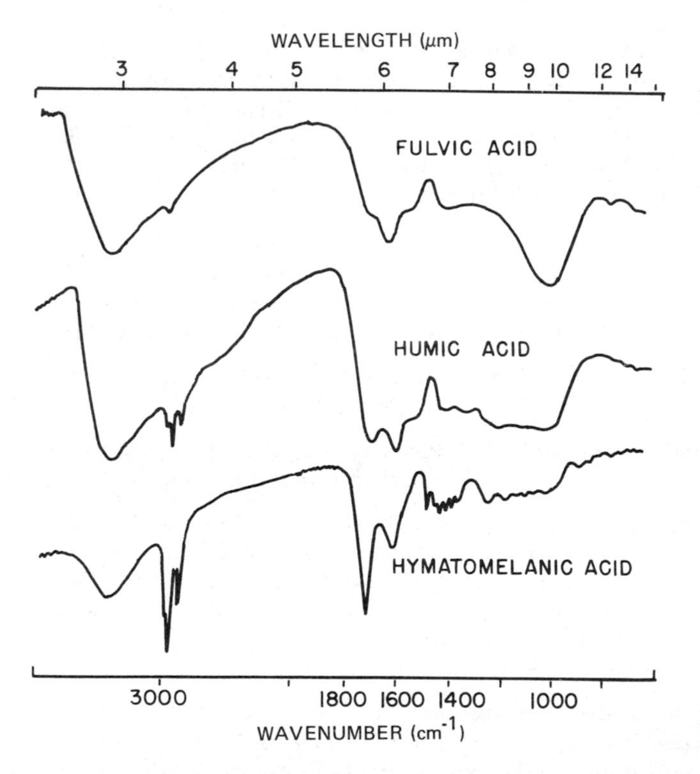

Figure 4.6 Diagnostic infrared characteristics of three major humic compounds, fulvic acid, humic acid, and hymatomelanic acid.

silica gel (Tan, 1976), but many other functional groups will also absorb in this region, such as ethyl, vinyl $-CH{=}CH_2$, aromatic aldehyde, amine, and SH groups. The infrared features of fulvic acid as discussed above have close similarities with those of polysaccharides (Mortenson, 1961).

In contrast to fulvic acid, humic acid exhibits a strong absorption for C—H vibrations at 2980 to 2920 cm^{-1}, and even stronger absorption for both carbonyl and carboxyl vibrations in COO^- form at 1720 and 1650 cm^{-1}, respectively. Humic acid spectra have in addition no absorption bands at 1000 cm^{-1}.

Hymatomelanic acid differs in turn from both fulvic and humic acids by showing infrared spectra with very strong absorption bands for C—H and carbonyl vibrations.

In addition to the functional groups discussed above for humic compounds, a number of other characteristic infrared group frequencies has been detected in organic compounds. This knowledge contributes toward solving the structural chemistry of humic acids, as will be discussed in a later section. The main absorption bands of humic compounds with their characteristic wave numbers and wave lengths are listed in Table 4.5.

Magnetic Resonance Spectroscopy. Two types of spectroscopy can be distinguished in this category, e.g., electron paramagnetic resonance, which analyzes electron spin resonance of large free radicals in large organic polymers of soil organic compounds, and nuclear magnetic resonance, used for determination of proton resonance in relatively smaller soil organic compounds. These methods have not found wide application because of the expensive machinery and the difficulty in using large complex humic compounds.

Electron paramagnetic resonance (EPR), also called electron spin resonance (ESR), analyzes unpaired electron spins in paramagnetic organic materials. The ESR spectra of humic and fulvic acids consist of single lines with hyperfine splitting and g values ranging from 2.0031 to 2.0045, with line widths from 2.0 to 3.6 G (Gosh and Schnitzer, 1980a; Riffaldi and Schnitzer, 1972). Eltantawy (1980) reported that an enhanced ESR signal (g = 2.0024) indicated the presence of larger free radical contents, while similar g values suggested that the free radicals in humic and fulvic acids were of similar structure.

Analysis of the EPR or ESR spectra by Steelink and Tollin (1967) and Steelink (1964) showed that humic acids exhibited paramagnetism due to the presence of semiquinones and hydroquinones. An EPR spectrum of a Spodosol (Podzol) and of humic acid extracted from the Spodosol is shown as an example in Figure 4.7. The peak in the Spodosol curve was identified as the organic

Table 4.5 Infrared Absorption Bands of Functional Groups in Humic Compounds and Related Organic Substances

Wavenumber, cm^{-1}	Wavelength		Proposed assignments
	μ	nm	
3400	2.94	2940	O—H and N—H stretch
3300	3.03	3030	O—H stretch and N—H stretch
3380	2.95	2950	Hydrogen bonded OH
3100	3.25	3250	NH_3 stretch
2985	3.35	3350	CH_3 and CH_2 stretch
2940	3.40	3400	C—H stretch
2920	3.42	3420	C—H stretch of CH_3
2900	3.44	3440	Aliphatic C—H stretch
2820	3.55	3550	CH_3 and CH_2 stretch
2740	3.65	3650	Hydrogen bonded OH
2610	3.83	3830	Hydrogen bonded OH
2260	4.42	4420	Hydrogen bonded OH
1840	5.43	5430	C=O stretch of cyclic anhydrides, and mixed anhydrides
1815	5.51	5510	C=O stretch of cyclic anhydrides, and mixed anhydrides
1785	5.60	5600	C=O stretch of phenols and cyclic anhydrides
1750	5.71	5710	COOH and C=O
1725	5.79	5790	C=O of carboxyl and C=O of ketonic carbonyl
1720	5.81	5810	C=O stretch of carbonyl groups
1695	5.90	5900	COOH vibrations
1680	5.95	5950	COO^- antisymmetric stretch

Table 4.5 (continued)

Wavenumber, cm^{-1}	Wavelength μ	Wavelength nm	Proposed assignments
1665	6.00	6000	Olefenic C=C
1650	6.0-6.2		C=O stretch (amide I)
1630	6.10	6100	Aromatic C=C, hydrogen bonded C=O, double bond conjugated with carbonyl and COO^- vibrations
1613	6.19	6190	C=C and COO^-
1610	6.20	6200	Conjugated C=C in ring with C=C, or C=O of open chains
1600	6.27	6270	Aromatic C=C
1590	6.29	6290	Multinuclear aromatic C=C and/or aromatic C=C
1575	6.35	6350	Salts of COOH
1570-1515	6.5-6.6		N-H deformation and C=N stretch (amide II)
1550	6.45	6450	COO^- antisymmetric stretch
1510	6.62	6620	Aromatic C=C
1470	6.80	6800	Aromatic C=C
1460	6.85	6850	$CC-H_3$
1440	6.95	6950	C—H stretch of methyl groups
1435	6.97	6970	C—H bending
1400	7.14	7140	COO^- antisymmetric stretch
1390	7.20	7200	Salts of COOH
1300	7.70	7700	C=N stretch and N—H deformation (amide III)

Table 4.5 (continued)

Wavenumber, cm^{-1}	Wavelength		Proposed assignments
	μ	nm	
1280	7.80	7800	C—O stretch
1267	7.89	7890	Aromatic C—O
1230	8.10	8100	C—O ester linkage and phenolic C—OH
1170-950	8.50-10.5	8500-10500	C—C, C—OH, C—O—C typical of glucosidic linkages, polymeric substances and Si—O impurities in humic compounds
1035	9.67	9670	O—CH_3 vibrations
840	11.9	11900	Aromatic C—H vibrations

Source: Adapted from Mortenson et al, (1965), Felbeck (1965), and Schnitzer and Khan (1972).

radical in the humic acid molecule. The spectrum of humic acid supported the observations above.

Schnitzer and Khan (1972) reported that if semiquinone is present, this compound can be made to enter in a reaction yielding hydroquinone and quinone:

O / OH — Semiquinone radical → OH / OH — Hydroquinone + O / O — Quinone

Hydroquinone + Quinone: Quinhydrone

The quinone released can be easily detected. However, chemical and spectroscopic analyses by the latter authors failed to show measurable quantities of quinone in solutions of degradation products of humic material. This lead Schnitzer and Khan (1972) to believe that evidence

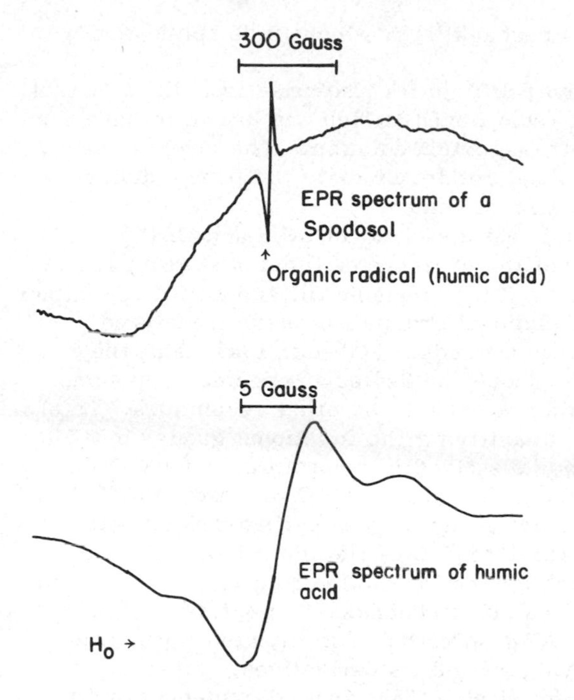

Figure 4.7 EPR spectra of a Spodosol (top) and humic acid extracted from the Spodosol. [From Steelink and Tollin (1967).]

presented by Steelink and Tollin (1967) for the presence of quinone groups in humic molecules was not convincing. However, Gosh and Schnitzer (1980a) recently confirmed by ESR analysis the presence of semiquinone radicals in humic substances. The free radical content in humic and fulvic acids was estimated to range from 1.4×10^{-17} to 37.4×10^{-17} spins/g (Gosh and Schnitzer, 1980a; Senesi and Schnitzer, 1977).

Nuclear magnetic resonance (NMR), the second method in the category of magnetic resonance spectroscopy, analyzes the hydrogen atoms, or proton resonance, of the humic molecules. Different types of protons in the unknown structure can be detected. However, the usefulness of NMR analysis in humic acid research is questioned by many authors. Nuclear magnetic resonance analysis requires the sample to be dissolved in a suitable solvent, the solvents frequently used were CCl_4 and $CDCl_3$. Humic acid, however, is usually not soluble in these reagents and must be first methylated or broken down into smaller molecules by degradation procedures. Solid samples cannot be used, since they interfere with magnetic interaction. Another solvent for NMR analysis is perhaps D_2O, which finds application with analysis of fulvic acids.

In addition to the above, difficulties also arise from the fact that NMR analysis makes use of radio waves, which are known to have the lowest-energy form of electromagnetic radiation. The level of energy involved is too small to vibrate, rotate, or exite the poorly defined complex polymers of humic acid molecules.

Using methylated fulvic acid dissolved in CCl_4 and $CDCl_3$, results of NMR analysis showed the presence of functional groups, such as aliphatic C—H, O—CH_3, CO_2CH_3, phenolic OH and COOH (Schnitzer and Khan, 1972). These functional groups can also be analyzed easily with infrared and other methods. Felbeck (1965) is of the opinion that the use of NMR should be limited to organic compounds that have been defined rather completely by other techniques. Lately, Ogner (1979) succeeded in quantifying the functional groups by NMR. Using methylated humic acids dissolved in chloroform and ^{13}C-NMR spectroscopy, Ogner estimated that 21% of the C was aromatic, 35% in methylene chains and in methyl groups, 3% as carbonyl-C of methyl esters, and 13% of the C as methoxyl-C of methylated phenols and alcohols. The remaining 28% was not accounted for; it was suggested to be mostly C bonded to O as polysaccharides or peptides. Ogner also indicated that the ^{13}C-NMR spectrum of methylated humic acid was different from those of lignin and its derivatives.

More recently, Ruggiero et al. (1980) showed evidence for the occurrence of exchangeable aromatic protons in fulvic and humic acids in their analysis using ^{1}H-NMR spectroscopy. According to these authors the protons in the aromatic structure of humic substances could be exchanged with deuterium. ^{1}H-NMR spectra enabled the proportion of exchangeable protons, expressed as percentage of total aromatic protons, to be measured. The estimated exchangeable proton concentrations were 18 ± 7% in humic acid, and 35 ± 10% in fulvic acid.

In addition to the above, NMR spectra can also serve as fingerprints of humic compounds. The present author used NMR (see Figure 4.8) to study adsorption of fulvic acids by soils. Fulvic acid, extracted from poultry litter, exhibits an NMR spectrum composed of at least four peaks (top curve in Figure 4.8). After shaking the fulvic acid solution with soil, the remaining supernatant was sampled and analyzed again. The NMR spectrum is shown at the bottom part of the figure. It can be noticed that peaks 1, 2, and 3 have decreased sharply in intensity, while peak 4 has disappeared. The latter suggests that the fulvic acid concentration in solution has been decreased, because of adsorption by soil.

Potentiometric Titrations of Humic Acids This method has been used extensively to characterize humic substances (Posner, 1964). The curves are presented usually as sigmoidal in nature, while titration

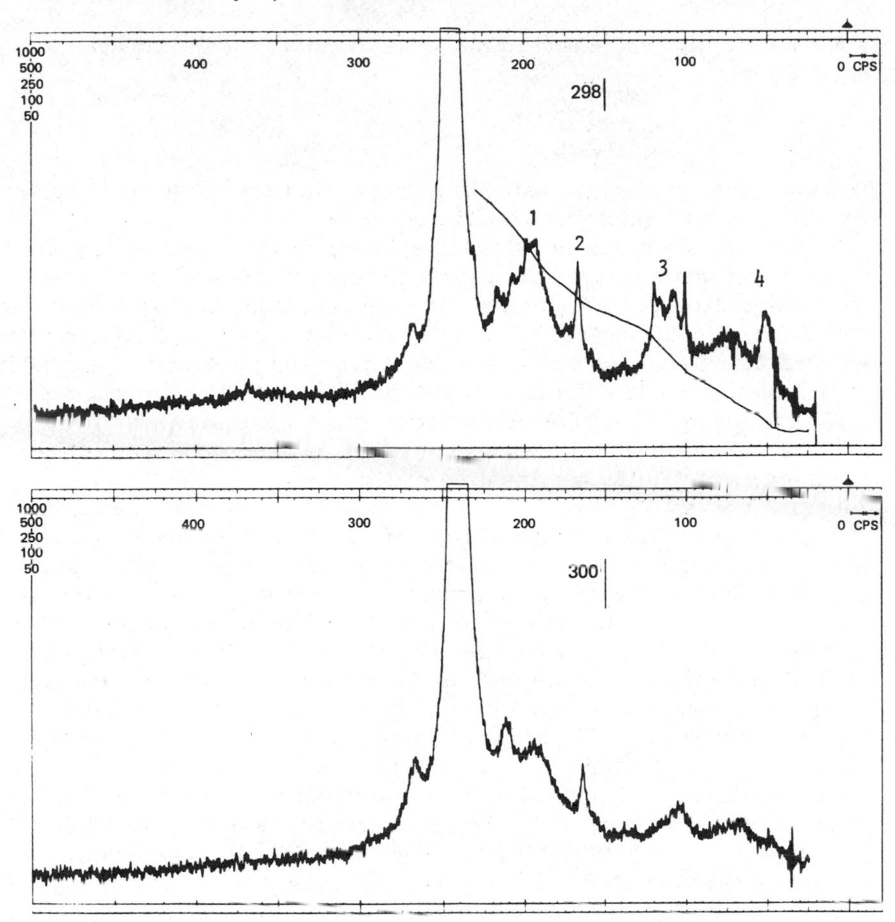

Figure 4.8 Nuclear magnetic resonance (NMR) analysis of fulvic acid extracted from chicken manure, before (top) and after (bottom) adsorption by Cecil (Ultisols) surface soil.

was carried out mostly with a base. This result leads to the belief that humic substances have a monobasic character (Schnitzer and Khan, 1972). Humic acids are in fact amphoteric and polybasic in nature. Depending on the soil condition they can be neutral, negatively charged, or positively charged. The negative charges are attributed to the presence of phenolic-OH and carboxyl groups. The dissociation of H^+ ions from these functional groups for the development of the negative charges can be illustrated with a carboxyl group as follows:

$$R-COOH \rightleftharpoons R-COO^- + H^+$$

for which the pK value according to the Henderson-Hasselbalch equation is

$$pK_1 = pH - \log \frac{(COO^-)}{(COOH)}$$

For pK values in the acid and basic range, and the corresponding titration curves, reference is made below.

Not only do humic acids possess two different types of oxygen containing functional groups, they also contain dissociable H^+ ions from amino groups. Although several investigators believe that humic substances should be free of N (Burges, 1960), elemental analysis revealing significant amounts of N (Table 4.1) suggests that N-containing compounds, such as amino acids, must be present in the humic molecule. As discussed earlier, amino acids are amphoteric in nature, and depending on conditions can behave as an acid or a base. The latter necessitates the continuation of titration in the acid range. Titration curves obtained by acid and base titration of humic acids are shown in Figure 4.9(a) and (b). The curves indicate that at pH 7.0 the carboxyl groups of humic acids are unprotonated, whereas the OH and/or amino groups appear to be protonated. The latter may produce substituted hydronium ions or substituted amino ions $(-NH_2-H)^+$. Addition of acid to the solution lowers the pH rapidly at first and then more slowly as the buffering action of the carboxyl groups is activated. At pH 2.4 the pK_a is reached, at which half of the carboxyl groups of humic acid (high molecular weight) are considered ionized, conforming to the Henderson-Hasselbalch equation. Further addition of acid results in a slight decrease in pH, and finally at pH 2.0, the high molecular weight humic acid starts to flocculate. Titration of the substituted OH and amino acids with base follows an opposite trend in the alkaline region. According to the inflection point, the pK_a is reached at pH 11.2.

Chromatography of Humic Substances Two types of chromatographic methods have been used in the study of humic material, e.g., gel chromatography and gas-liquid chromatography, for quite different purposes. Gel chromatography is proposed for purification processes or for fractionation of the humic compounds into different components or into fractions of different molecular sizes. The latter has been extended to measurements of molecular weights (Orlov, et al., 1975; Holty and Heilman, 1971; Swift et al., 1970; Mehta et al., 1963). On the other hand, gas-liquid chromatography is used primarily to fractionate the humic material into its component molecules. However, both methods can also be applied in the characterization of humic substances by using the chromatograms as fingerprints. Unfortunately the advantages of the latter have not been fully exploited. Today the use of gas chromatography together with mass spectrometry

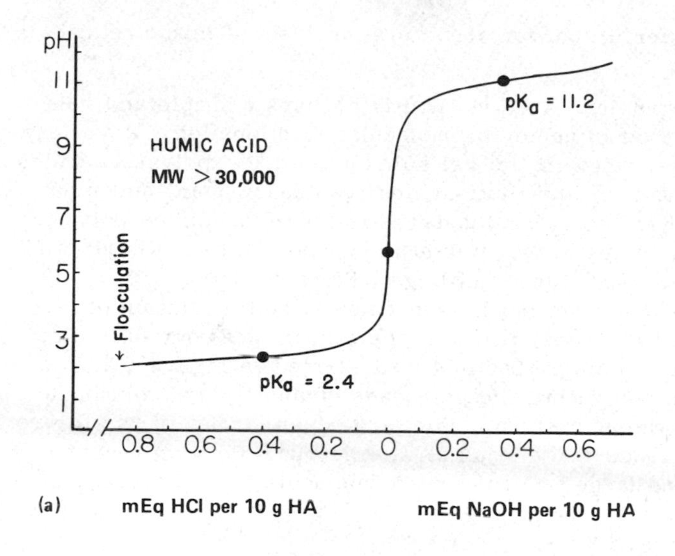

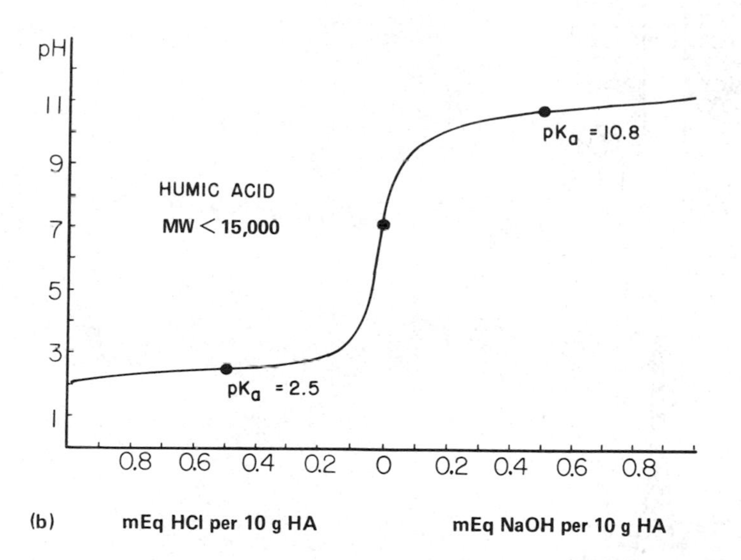

Figure 4.9 Titration curves of (a) high (>30,000) and (b) low (<15,000) molecular weight fractions of humic acid from a Cecil soil (Ultisols). Molecular fractionation was done by gel chromatography using Sephadex G-50. [From Tan and Giddens (1972).]

is considered a powerful tool for structural analysis of humic acids (Schnitzer, 1976).

Gel Chromatography. This method is perhaps a simple and relatively effective method in achieving molecular fractionation. A variety of materials has been used as the gel substance, e.g., polysaccharides, polystyrene, polyamides, aluminum oxides, cellulose, agar, and glass beads. The most widely used gel material is the cross-linked polymer of polysaccharides, polysterene, polyamides, and the like, distributed under names such as Sephadex, Sephagel, Bio-gel, Cellogel.

A column of swollen gel beads is prepared, and a solution of humic acid is eluted (filtered) through the column at a controlled flow rate. The filtration is diagrammatically illustrated in Figure 4.10. The pores between and within the gel beads enable the gel column to act as a chromatographic medium. The large open circles in the figure are the Sephadex beads. The small and large black dots represent a mixture of small and large molecules of humic acids. As the mixture

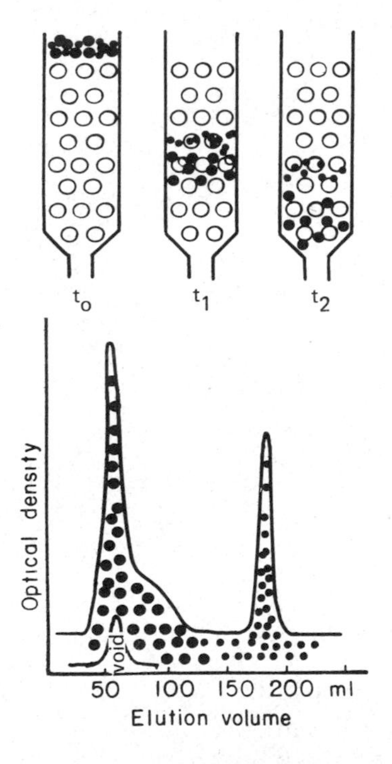

Figure 4.10 Sephadex G-50 gel filtration of humic acid from a Cecil soil (Ultisols): (●) high molecular weight HA, (•) low molecular weight HA, (O) Sephadex beads.

of molecules passes through the column, the larger molecules are eluted first and depending on conditions, the elution curve may be represented by two peaks, as shown in the figure. The first peak is attributed to the larger molecules, and the second peak to the smaller molecules. If the mixture is composed of only small molecules, such as fulvic acids, the elution curve may be characterized by only one major peak (see Figure 4.11).

The behavior of the solute in gel filtration can be measured by its elution volume, while the elution peak is determined by the partition coefficients K_{av} and K_d, which are defined as

$$K_{av} = \frac{V_e - V_o}{V_t - V_o} \qquad K_d = \frac{V_e - V_o}{V_i}$$

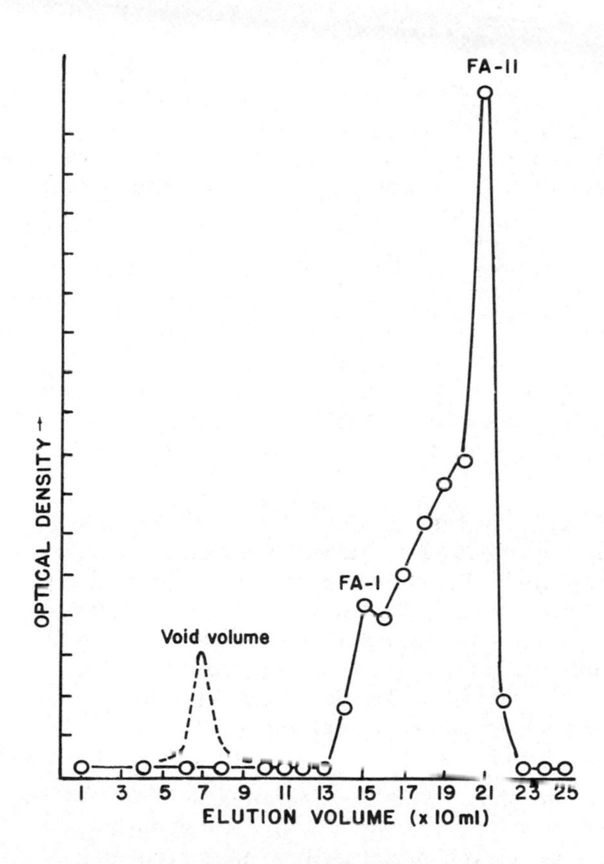

Figure 4.11 Gel filtration of fulvic acid using Sephadex G-50. FA-I has larger size molecules than FA-II.

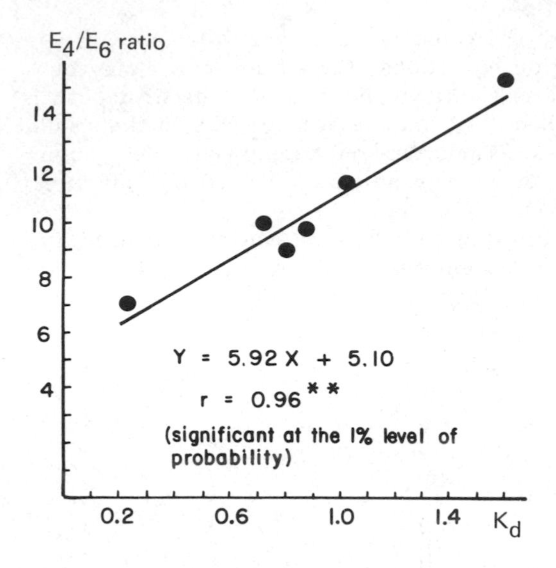

Figure 4.12 The relationship of color ratios (E_4/E_6 ratios) of fulvic acids and partition coefficients (K_d values). [From Tan and Giddens (1972).]

where

V_e = elution volume
V_o = void volume
V_t = total volume of gel bed
V_i = inner volume of the gel (Handbook Sephadex-gel filtration in Theory and Practice, Pharmacia Fine Chemicals, Uppsala, Sweden, (1969).

Tan and Giddens (1972) reported that the partition coefficients K_d were higher for the small molecules of humic material, and a positive correlation was obtained between K_d values of humic fractions separated with Sephadex and color ratios (E_4/E_6 ratios). As can be noticed from Figure 4.12, humic fractions with higher partition coefficients exhibit larger E_4/E_6 ratios. This corresponds to the observation indicating that the smaller molecules eluted last also have lower molecular weights, whereas the larger molecules, eluted first, are of higher molecular weights.

Gas-Liquid Chromatography. This is a complex method not as versatile as gel chromatography. Gas chromatography requires the material to be analyzed to be easily transferred into the gaseous phase, which is a major obstacle with humic acids. Humic compounds are generally nonvolatile, and no method has been reported thus far in the literature for making them volatile without decomposing the humic material into smaller components first. Schnitzer and Khan (1972) reported the senior author's efforts in gas chromatography using degradation products of fulvic acids, obtained by oxidation with strong acids or by reduction with sodium amalgam. The degradation products identified were mostly benzene derivatives, e.g., benzene di-, tri-, tetra-, penta-, and hexacarboxylic acids. Although valuable information can thus be obtained, much remains unknown concerning the natural assembly patterns in humic acids. The extent to which organic artifacts that no longer resemble the real structural units of the original humic material may be produced during the oxidation or reduction processes is not clear.

Success has been obtained in gas chromatography of carbohydrates and related polyhydroxy compounds with the use of TMS, trimethylsilyl (Tan and McCreery, 1970a). The TMS group $Si(CH_3)_3$, was then introduced (by shaking) into humic substances through their functional groups, which have exposed protons, or active H^+ ions:

$$R-COOH + TMS-Cl \rightarrow R-COO-TMS + HCl$$

Silylated humic acid from different kinds of soil exhibited gas chromatograms (see Figure 4.13) with almost the same number of components appearing at similar retention times (Tan and McCreery, 1970b). However, the concentration varied considerably between some of the components of the two humic acids, as can be seen by the differences in peak height.

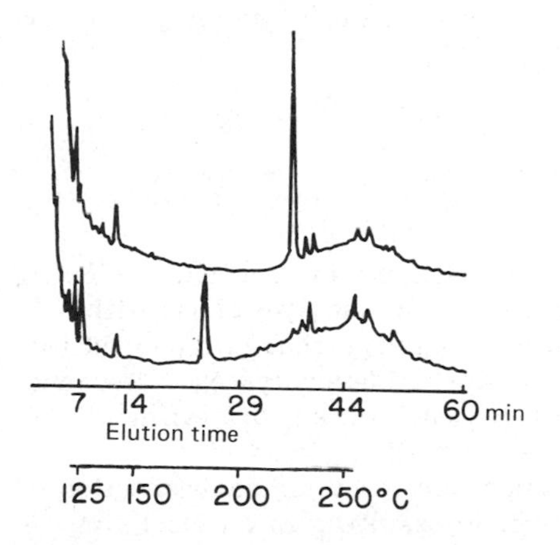

Figure 4.13 Gas chromatograms of humic acids silylated for 36 hr. Methylated HA from a Cecil soil (Ultisols, top). Methylated HA from a Greenville soil (Ultisols, bottom). [From Tan and McCreery (1970b).]

Molecular Weights of Humic Acids The use of molecular weights in characterization of humic substances encounters many problems, because these compounds are known to be polydispersive. They possess, therefore, a wide spread in molecular weights, causing the humic material to be very heterogeneous in molecular weights (Felbeck, 1965). The degree of polydispersity may vary considerably depending on the contributing components of the humic molecule with different molecular weights.

Physically, molecular weights can be expressed as follows:

1. The *number-average molecular weight* $\overline{M}_n$ is defined as

$$\overline{M}_n = \frac{\Sigma\, nM}{n}$$

where

n = number of component molecules
M = molecular weight of component molecules

The methods used to determine $\overline{M}_n$ are osmometry, diffusion, isothermal, and cryoscopic distillation. Osmometry is considered the best method, but it appears not to be applicable to analysis of molecular weights >200,000.

2. The *weight-average molecular weight* $\overline{M}_w$ is defined as

$$\overline{M}_w = \frac{\Sigma\, nM^2}{\Sigma\, nM}$$

The weight-average molecular weight is usually measured using viscosity analysis and gel filtration. Of the two, gel filtration is the most simplest method as discussed earlier.

3. The *Z-average molecular weight* $\overline{M}_z$ which is defined as:

$$\overline{M}_z = \frac{\Sigma\, nM^3}{\Sigma\, nM^2}$$

The method used to measure $\overline{M}_z$ is the sedimentation method employing the ultracentrifuge. It is considered to yield many problems with humic compounds. Humic acid carries a negative charge balanced by cations, creating in this way a diffuse double-layer system. Because of the latter, the molecules tend to repel each other, offsetting the sedimentation process. Intermolecular repulsion yields high-diffusion and low-sedimentation coefficient values due to faster sedimentation of the larger molecules than the counterions resulting in an electrostatic drag. In addition, the polydisperse nature makes it difficult to achieve well-defined sediment boundaries with humic material.

For a heterogeneous, or polydisperse, system, $\overline{M}_n < \overline{M}_w < \overline{M}_z$, but for a homogeneous, or monodisperse, system, $\overline{M}_n = \overline{M}_w = \overline{M}_z$.

Values reported for weight-average molecular weight M_w of humic substances may vary from 1000 to 30,000. Flaig and Beutelspacher (1951) stated molecular weights of >100,000, and values of 2,000,000 have been reported occasionally. Apparently any number within these ranges can be obtained depending on the chemical isolation procedures and analysis employed, with usually fulvic acids exhibiting the lower and humic acids the higher molecular weight values. Tan and Giddens (1972) reported that with Sephadex G-50 gel filtration, humic acid was separated only into two fractions, e.g., a high (>30,000) and a low (< 15,000) molecular weight fraction (see Figure 4.14). The high molecular weight fraction made up 50% of the humic acid isolate, while the remaining 50% was the low molecular weight humic compound. Using the same gel filtration technique, Tan and McCreery (1975) noted that the degree of polymerization or the size of the molecules isolated affected molecular weight. The data in Table 4.6 demonstrate the relationship of size of molecule and molecular weight. By assuming the humic molecules to be spherical in shape, the larger the size of the molecules of the humic fraction isolated, the larger will be the numerical value of the molecular weight of humic acid.

Electron Microscopy of Humic Acids Flaig and Beutelspacher (1951) were perhaps among the first who tried electron microscopy in the study of humic acids. This method, enjoying considerable success in clay mineralogy, was used to analyze the morphology and dimension of humic particles. Employing the transmission electron microscope, Flaig and Beutelspacher indicated that humic acids occurred as very small spherical particles, of the order of 100 to 150 Å in diameter. The spheres frequently joined together in racemic chains. Since 1951 very few investigations have been carried out using this method (Schnitzer and Khan, 1972).

Recently, scanning electron microscopy (SEM) has shown better possibilities and greater promise for investigation of shape, size, and degree of aggregation of fulvic and humic acids and their complexes (Chen and Schnitzer, 1976; Dormaar, 1974). This method has the advantage over the standard transmission electron microscopy since it furnishes a three-dimensional picture with a resolution depth of 5 to 10 μm. In addition, humic particle surfaces and orientation can also be shown. The method has not been fully exploited, and procedures of sample preparation to avoid altering or damaging the original structure of humic particles have not yet been completely established. Chen and Schnitzer (1976) used ultrarapid freezing techniques for their SEM analysis. This has been adapted from preparation of animal tissue for surface scanning electron microscopy. Their SEM micrographs (see Figure 4.15) showed that protonated fulvic acid (adjusted to pH 2 to 3) exhibited an open structure, formed from elongated fulvic acid strands or fibers with rounded or spherical tips. This structure

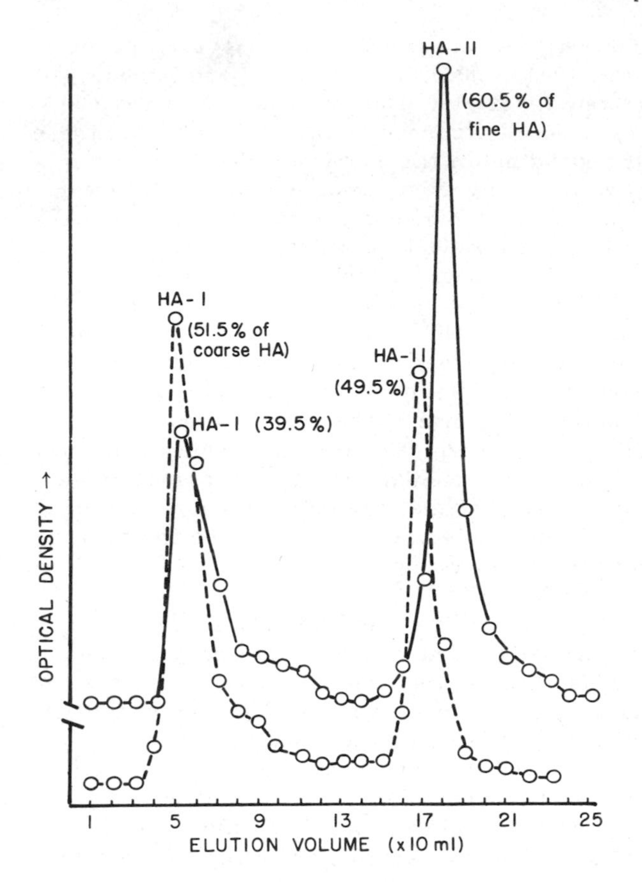

Figure 4.14 Sephadex G-50 gel filtration of humic acids. HA-I = high-molecular-weight fraction of humic acid; HA-II = low-molecular-weight fraction of humic acid. Coarse HA (broken line) means dialysis of humic acid conducted with membrane with coarse pores (molecular weight cutoff of membrane is 10,000). Fine HA (solid line) means dialysis of humic acid conducted with cellulose membrane with fine pores (molecular weight cutoff of the membrane is 3500).

changed into a spongelike structure if the pH of fulvic acid was adjusted to 7.0, while at pH 9.0 fulvic acid revealed homogeneous sheets, in which grains were visible. The behavior of macromolecular structures of humic acids appeared to vary not only according to pH, but also according to ionic strength and sample concentration (Gosh and Schnitzer, 1980c). At high concentration, low pH or at medium to high ionic strength, they behaved as rigid spherocolloids but were otherwise flexible linear colloids.

Table 4.6 Molecular Weights and Size (in Å and nm) of Humic Fractions Obtained by Sephadex Gel Filtration

Molecular weight	Molecular volume, $Å^3$	Radius	
		Å	nm
30,000	23,622	17.8	1.78
5,000	3,937	9.8	0.98
1,500	1,181	6.6	0.66
1,000	787	5.7	0.57

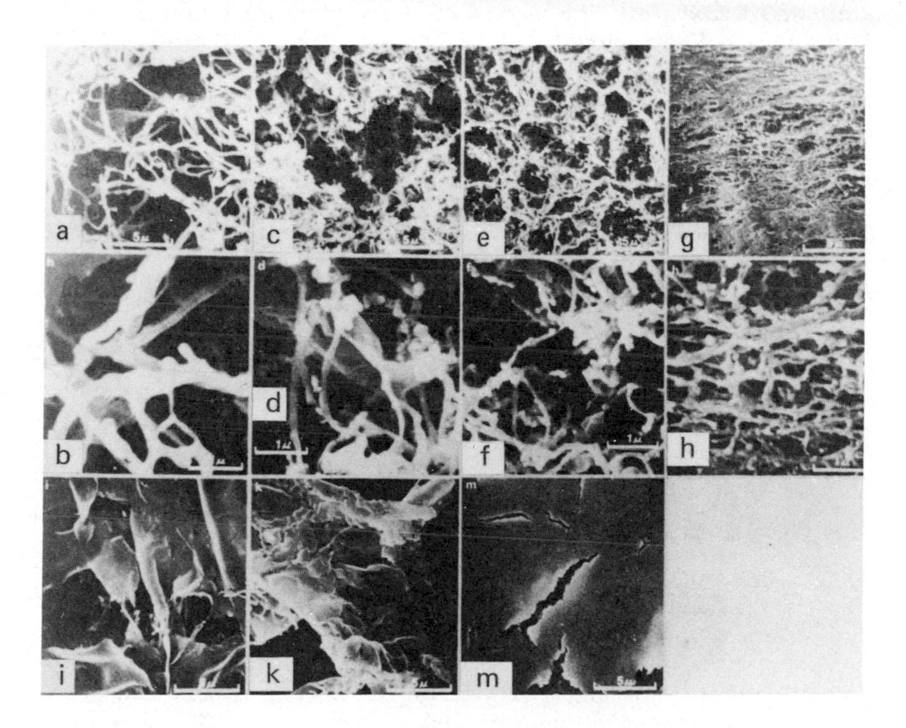

Figure 4.15 Scanning electron micrographs of fulvic acids at various pH values: (a) and (b) at pH 2.0, (c) and (d) at pH 4.0, (e) and (f) at pH 6.0, (g) and (h) at pH 7.0, (i) at pH 8.0, (k) at pH 9.0, (m) at pH 10.0. [From Chen and Schnitzer (1976). Reproduced by permission of the Soil Science Society of America.]

Structural Chemistry of Humic Acids A number of hypotheses have been reported on the structural chemistry of humic acids, but apparently the hypotheses lack desirable uniformity and a number of disagreements still exist.

Hypothesis of Schnitzer and Khan (1972). The concept is based on information obtained from chemical degradation of fulvic acids. Degradation reactions do not ensure that artifacts may not have been produced. Depending on the severeness or mildness of the reactions applied, any breakdown product can be obtained ranging from elemental C, H, O, benzene rings to heterocyclic rings. Schnitzer and Khan are of the opinion that humic substances must be broken down into smaller subunits in order to study their structural chemistry. Thus far four basic types of degradation procedures have been used:

1. Oxidation with alkaline permanganate, nitric acid, H_2O_2, and CuO-NaOH mixture. The degradation products were invariably benzenecarboxylic acids.

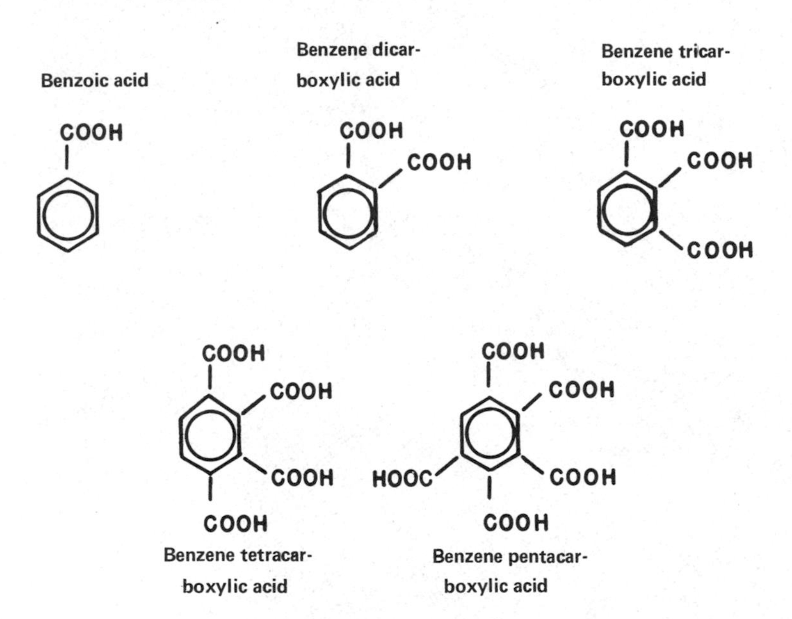

2. Reduction with Na amalgam or with Zn dust. Fulvic acid also yielded benzene derivatives with this method.
3. Hydrolysis with hot water, with acids or bases. Fulvic acid yielded benzene derivatives such as hydroxy benzoic and vanillic acids.
4. Biological degradation. This is a method by which fulvic acid is decomposed with the aid of microorganisms, e.g., *Penicillum* sp.,

Aspergillus sp., *Trichodermia* sp. The compounds produced from fulvic acid by biological degradation were also identified as benzene derivatives.

On the basis of the predominant findings of benzene derivatives, Schnitzer and Khan (1972) assume that fulvic acid is composed of phenolic and benzene carboxylic acids, joined together by hydrogen bonds to form a polymeric structure. The latter contains many voids or openings in which other organic compounds, such as amino acids and carbohydrates, can be trapped.

Hypothesis of Kononova (1961). Kononova is of the opinion that at least three basic steps are involved in formation of humic acids: formation of structural units from the decomposition of plant tissues, condensation of these units, and polymerization of the condensation products. The result is a multicomponent system, called humic or fulvic acids. They show similar structural patterns but may differ in details of structural and chemical composition, e.g., fulvic acid has a less condensed aromatic nucleus, but has more highly developed pheripheral components. In Kononova's opinion fulvic acid can be both the predecessor or the decomposition product of humic acid.

The basic structural units of humic compounds are considered to be phenolic or quinoid in nature, bonded to nitrogen-containing compounds and carbohydrates, the latter chiefly polyuronides. The inclusion of N-containing compounds and carbohydrates as fundamental units of the humic molecule is a matter of many controversies. Several investigators regard the latter as accidental contaminants trapped in the mazework of the humic structure (Burges, 1960; Schnitzer and Khan, 1972), but others show evidence for the necessary participation of carbohydrates and N compounds in the formation of humic acids (Kononova, 1961; Flaig et al., 1975). Kononova (1961) suggests the following reaction to occur for the inclusion of N in the humic molecule:

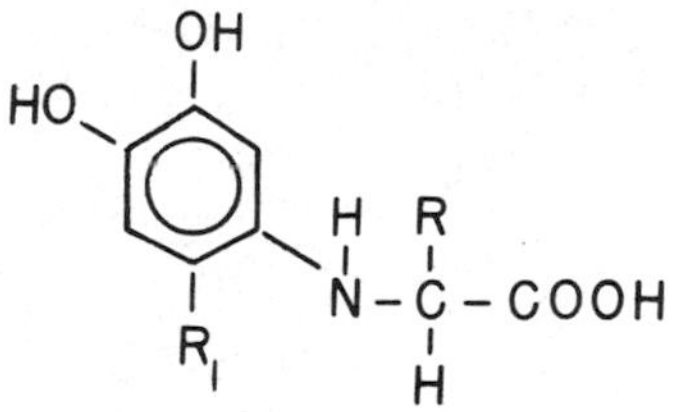

OH
HO
H R
N-C-COOH
R_1 H

Such a combination produces a stable condensation product of phenols and α-amino acids and increases the stability of N with respect to acid hydrolysis.

As indicated above, a number of authors disagree on including carbohydrates in the humic molecule; they are of the opinion that they are contaminants that can be removed by adequate purification methods. Notwithstanding these arguments, various research reports (Clark and Tan, 1969; Tan and Clark, 1968; Tan, 1975) point to the discovery of polysaccharides as an integral part of the humic molecule. Fulvic acid appears to be composed of large amounts of polysaccharides, while hymatomelanic acid has been identified as a polysaccharide ester compound. Infrared spectroscopy of hymatomelanic acid shows a characteristic strong absorption between 3000 and 2800 and at 1720 cm^{-1}, the latter usually accompanied by a weak shoulder at 1620 cm^{-1} (see Figure 4.16).

These absorption features can be reproduced artifically, or closely simulated, by methylation of humic acid. In addition, the in-

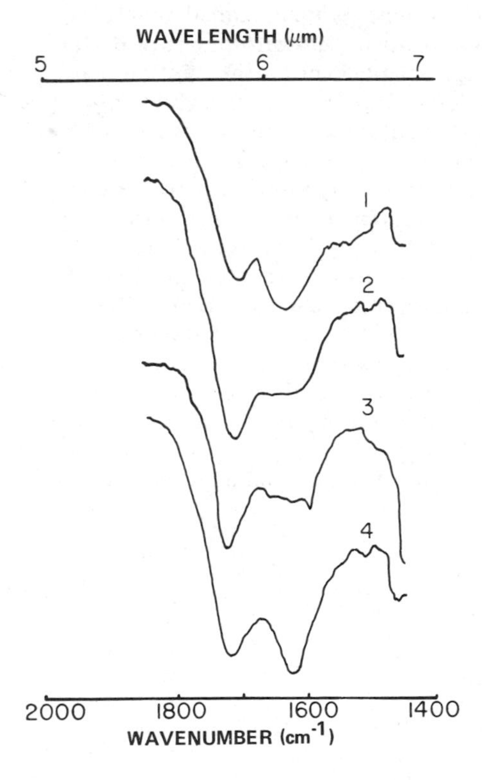

Figure 4.16 Infrared spectra of humic fraction between 2000 and 1400 cm^{-1}, showing absorption attributed to carbonyl (1720 cm^{-1}) and carboxyl (COO^-, 1625 cm^{-1}) groups. (1) Purified humic acid, (2) hymatomelanic acid, (3) methylated humic acid and (4) hymatomelanic acid, after removal of ester group. [From Tan and McCreery (1970b).]

frared absorption of hymatomelanic acid at 1620 cm^{-1} can be increased considerably in intensity by separation of the ester group through hydrolysis of the compounds with 2 N NaOH. The latter procedure removes the blocking effect of the ester group on the carboxyl vibration in COO^- form. Identification of the ester fraction reveals infrared features similar to those of a soil polysaccharide (Mortenson, 1961).

Ogner (1980) indicated that polysaccharides in humic compounds were highly branched and complex in composition. They contained at least 34 different aldose building units, in addition to O-methyl, di-O-methyl, amino derivatives, ketones, and uronic acids. Generally, polysaccharides of fulvic acids were comparatively less branched than those of humic acids.

Hypothesis of Flaig. Flaig and co-workers (Flaig et al., 1975) suggest lignin to be the source, or starting point, for the formation of humic and fulvic acids. Lignin is assumed to be broken down by degradation or decomposition reactions into its basic units, e.g., coniferyl alcohol or guaiacyl propane monomers. These lignin basic units are then subject to oxidation, followed by demethylation to substituted polyphenols and further oxidation to quinone derivatives. Condensation of the quinone groups with amino acids and polysaccharides may then yield humic acid-like substances. Lignin degradation products have been detected in hydrolyzates of humic acids.

Lignin → Oxid. → Demeth. → Oxid.

3-5-Dimethoxy-
4-hydroxy
phenyl propane

Nucleus for
humic acids

Effect of Humic Acids on Plant Growth The effect of organic matter on plant growth has been known for some time. The major benefits of soil humus on plant growth result indirectly through improvement of soil properties such as aggregation, aeration, permeability, and water-holding capacity. With the increased knowledge of soil organic matter, and in particular of humic acid chemistry starting a decade ago, increasing amounts of information have been accumulated on the direct influence of soil organic matter and humic acids on plant growth. Unconfirmed reports also mention that nucleic acids, protein, and small degradation products of humic acids can be taken up by plants (Flaig, 1975). Organic compounds such as acetic, propionic, butyric and valeric acids have been noted to increase root growth only when they were present in combination. Alone, they had no effect, as studied by Wallace and Whitehand (1980) with root elongation of germinating wheat on agar media. In summary, it can be stated that humic acid compounds and the like can improve plant growth directly by acceleration of respiratory processes, by increasing cell permeability, or by hormone growth action. Most of the investigations with humic acids, limited to studying seed germination, shoot growth, and elongation of very young seedlings, or root elongation of excised roots in vitro, show the presence of a hormonal growth effect (Poapst et al., 1970, 1971). On the other hand, work done on nutrient uptake by Guminski and others in East Europe (Guminski et al., 1977; Guminski, 1957) and by Dormaar (1975) in Canada reveals the physiological influence of humic acids on plant growth. From studies on the growth and nutrient uptake of corn plants (*Zea mays* L.), Tan and Nopamornbodi (1979) came to the conclusion that humic acid affected plant growth through a combination of the processes mentioned above. As can be noticed from Figure 4.17, moderate amounts of humic acid were generally beneficial to root and shoot growth of corn plants. At the same time a significant increase in N content of shoots of corn seedlings was obtained. Dry matter production also appeared to be stimulated by moderate amounts of humic acids.

4.3 THE INORGANIC COMPONENTS

Clay Fraction of Soils

The inorganic fraction of soils is composed of rock fragments and minerals of varying size and composition. Despite the variability in composition, the inorganic fractions are predominantly silicates and oxides. They are sometimes distinguished into primary and secondary minerals. But this distinction creates problems, since secondary mineral deposits may well be regarded as primary on a pedological basis. On the basis of size three major fractions are usually recognized: (1) the coarse fraction (2 to 0.050 mm) called *sand*, (2) the fine fraction (0.050 to 0.002 mm) called *silt*, and (3) the very fine fraction (<0.002 mm) referred to as *clay* (USDA, 1975). In soil science we are used to considering clay as a colloid, although strictly speaking only the clay fraction <0.2 μm is colloidal clay.

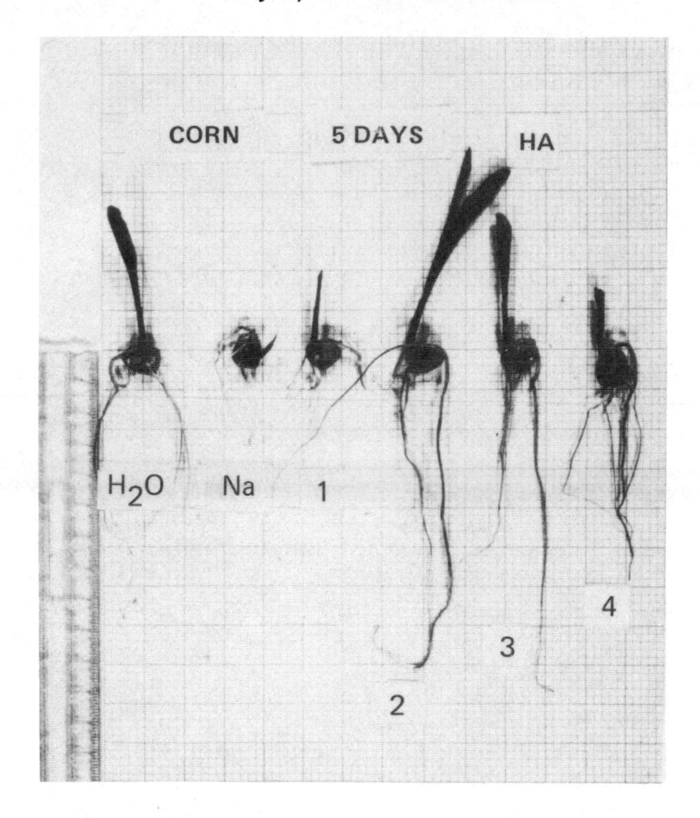

Figure 4.17 Germination and growth of 5-day-old corn seedlings as affected by humic acid: (H_2O) 0 ppm HA, (Na) blank + 0.66 mEq NaOH, (1) 320 ppm HA, (2) 640 ppm HA, (3) 1600 ppm HA, (4) 3200 ppm HA. The corn was grown in a modified Hoagland solution to which the HA treatments were applied. [From Tan (1978) and Tan and Nopamornbodi (1979).]

Six types of soil silicates are usually recognized on the basis of the arrangement of the SiO_4 tetrahedra in their structure:

1. Cyclosilicates	Closed rings or double rings of tetrahedra (SiO_3, Si_2O_5)
2. Inosilicates	Single or double chains of tetrahedra (SiO_3, Si_4O_{11})
3. Nesosilicates	Separate SiO_4 tetrahedra
4. Phyllosilicates	Sheets of tetrahedra (Si_2O_5)
5. Sorosilicates	Two or more linked tetrahedra (Si_2O_7, Si_5O_{16})
6. Tectosilicates	Framework of tetrahedra (SiO_2)

Table 4.7 Examples of Mineral Species Classified According to the Six Types of Soil Silicate

Soil silicate	Mineral species
Cyclosilicates	Tourmaline
Inosilicates	Amphibole, pyroxene, hornblende
Nesosilicates	Garnet, olivine, zircon, topaz
Phyllosilicates	Chlorite, illite, kaolinite, montmorillonite, vermiculite
Sorosilicates	Epidote
Tectosilicates	Feldspars, zeolite

Examples of mineral species belonging to the respective categories above are listed in Table 4.7. The sand and a major part of the silt fraction are cyclo-, ino-, meso- soro-, or tectosilicates. They make up the framework of the soil. Since they are coarse in size, they have low specific surface area and do not exhibit colloidal properties. Although not really active in chemical reactions, they participate in a number of reactions and exhibit some adsorption. Many of the sand and the silt minerals are also of importance for formation of clays. Some of the clays are phyllosilicates.

Clays are of special importance in soil chemistry, since they have a different surface chemistry from that of the larger mineral grains. Clays also exhibit bulk physical properties different from gravel, sand, or silt. Many of the minerals in soil clays are crystalline in structure, whereas others may poorly exhibit crystals or be structurally disordered. Some of the clays may be amorphous, e.g., silica, alumina, and iron oxide gels. The latter may occur in soils as discrete, independent, minerals, and/or as coatings around crystalline clay particles and other inorganic soil constituents. As stated above, not all of the clays belong to the phyllosilicate group, or layer-lattice silicates. The soils clay fraction also contain other minerals such as the palygorskite-sepiolite minerals, which are chain-structured minerals, quartz in particle sizes of <2 μm, sesquioxides, titanium oxides, pyrophyllite, talc, sulfides, sulfates, and phosphates. The major types of phyllosilicates are listed in Table 4.8.

Structural Chemistry of Clay Minerals

As indicated earlier, soil clays can exist in crystalline, structurally disordered, or amorphous form. The amorphous state generally has no recognizable shape or geometrical internal arrangement of atoms. Depending on degree of sophistication in methods for analysis, a sharp distinction does not exist between crystalline and amorphous states. In soil science, clay is considered amorphous if it is amorphous

Table 4.8 Major Phyllosilicate Minerals in Soils

Layer type	Group name	Charge per unit formula	Common minerals
1:1	Kaolinite-serpentine	∿ 0	Kaolinite, halloysite Chrysotile, lizardite, antigorite
2:1	Pyrophyllite-talc	∿ 0	Pyrophyllite and talc
	Smectite, or montmorillonite-saponite	0.25-0.6	Montmorillonite, beidellite, nontronite Saponite, hectorite, sauconite
	Mica	∿ 1	Muscovite, paragonite Biotite, phlogopite
	Brittle mica	∿ 2	Margarite, clintonite
	Illite	2	Illite
	Vermicullite	0.6-1.9	Vermicullite
2:1:1	Chlorite	variable	Chlorite
Chain	Palygorskite-sepiolite	—	Palygorskite, sepiolite

Source: Mackenzie (1975) and Brindley et al. (1968).

to x-ray diffraction analysis, i.e., lacks regularity in internal atomic arrangement as reflected by a featureless diffractogram. In crystals, the atomic arrangement may be repeated in a regular pattern in three-dimensional direction. However, in amorphous materials, such as glass, the chemical bonding of the component atoms is, perhaps, the only unit repeating itself. The spatial arrangement of atoms producing the building unit of a crystal is called the *unit cell.* The latter exhibits a complete group pattern of atoms that repeats itself in three directions in space according to the so called x, y, and z axes, respectively (see Figure 4.18). The z axis is sometimes called the c axis. The size or length of the edges of the unit cell in each direction is expressed in terms of a, b, and c, each of which represents a discrete unit length for a specific crystal, measured along the x, y, and z axes. In a cubic crystal, a, b, and c are of equal length, and the angles, α, β, and γ are 90°. In clay minerals, these angles may vary according to the structure. By placing several unit cells to-

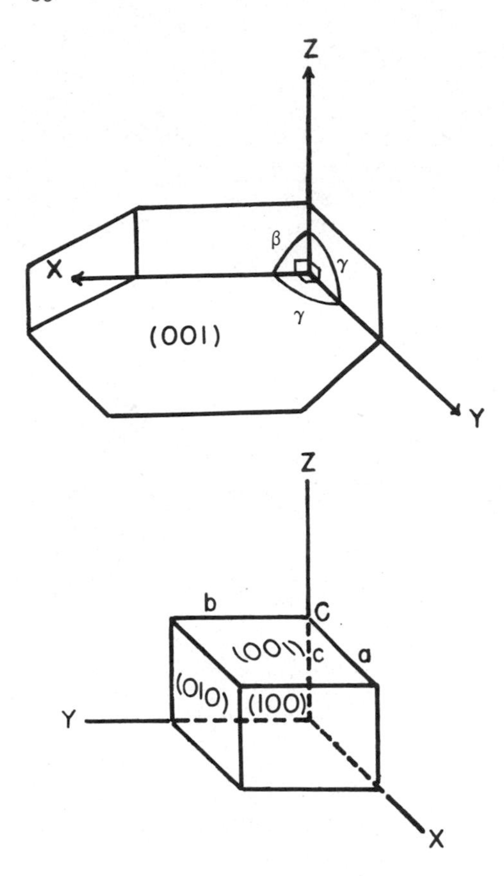

Figure 4.18 (Top) Crystal structure, exhibiting a group pattern of atoms, repeating itself in three directions in space, according to the so-called x, y, and z axes, respectively. (Bottom) Close-up of a unit cell, showing discrete unit length a, b, and c, measured along the x, y, and z axes in a cubic crystal.

gether, the crystal arrangement produced is then called a *lattice structure.* A perfect crystal may be composed of unit cells, each of which has a volume of approximately 1 μm^3.

The atomic groups in a crystal lattice can be arranged in planes at equal spacings along the crystallographic direction. Several types of atomic planes can be drawn in the crystal with interplanar spacings called *d spacings*. The plane delineated, or bordered, by a and b parallel to the x and y axes (see Figure 4.18) cuts the z axis at C but does not cut the x and y axes. According to the Miller indices system (Grimshaw, 1971), this plane is given the number 001. The basal (001) spacing plays a fundamental role in identification of clay mineral species by x-ray diffraction analysis.

Silicates are built around a silica tetrahedron, in which each oxygen atom receives one valency from the silicon atom. To satisfy its divalent requirement, the oxygen atoms can be linked to other cations or to a silicon atom of an adjacent silica-tetrahedron (see Figure 4.19). The linkage of silica tetrahedra yields five groups of structural arrangements of silicates: island, isolated group, chain, sheet, and framework structure. Silicate clay minerals are characterized by a sheet structure. In contrast to the other silicates, the structure of clay is not a three-dimensional network of simple linkages of silicon-oxygen units, but it is build up of stacked layers of silica tetrahedra and Al (Mg) octahedra sheets. The sheets are developed by the linkage of three oxygens in each tetrahedron with adjacent silica tetrahedra units as discussed earlier. The silica tetrahedra are arranged in hexagonal rings

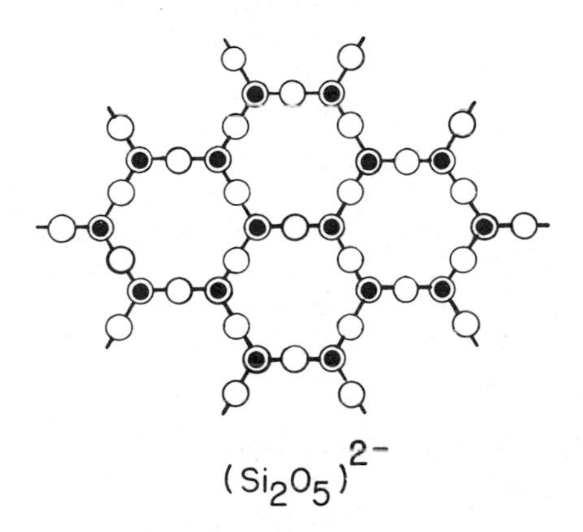

and the sheet can extend indefinitely in a two-dimensional direction, according to the two planes a and b, or parallel to the plane of the paper in this book, which is the reason for the platelike nature of clays. The composition of each ring, or the lowest unit of the sheet is

$$(Si_2O_5)^{-2} = \frac{6\ Si}{3} + \frac{6\ O}{2} + \frac{(6\ O^-)}{3}$$

In such a network of silica tetrahedra, one oxygen in each tetrahedron remains electrically unbalanced. To satisfy the divalent requirement, the latter is linked to Al in octahedral coordination. By such a packing of silica tetrahedron and aluminum octahedron sheets, a layered clay structure is formed. Several layers of silica tetrahedron and aluminum octahedron sheets can be stacked one above another. However, each layer is an independent unit and is considered the

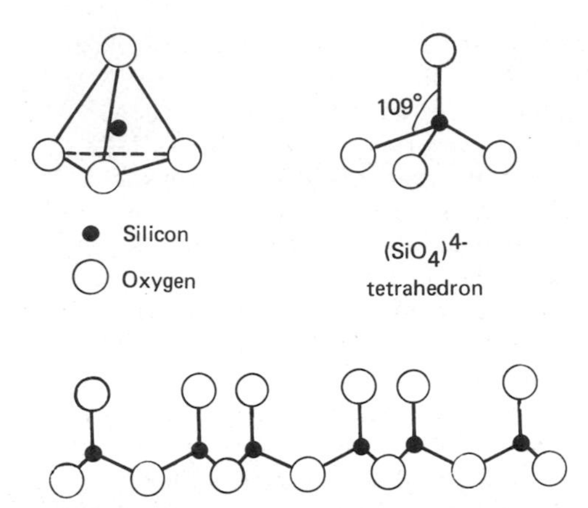

Figure 4.19 (Top) Schematic structure of a single silica tetrahedron. (Bottom) The arrangement of several silica tetrahedra into a sheet by mutually sharing oxygen atoms.

crystal unit. The bonds between the layers can be relatively strong, as in kaolinite, or can be relatively weak, as in montmorillonite. Within each layer a certain atomic grouping repeats itself in lateral direction. This group or unit layer is referred to as the *unit cell*, while the total assembly of a layer plus interlayer material is called a *unit structure*. Illustrations of the packing of silica tetrahedra and aluminum octahedra will be given in the sections on the major types of clay minerals.

On the basis of the number of tetrahedral to octahedral sheets in one layer, the following structural types are recognized: 1:1 or dimorphic, 2:1 or trimorphic, 2:2 or 2:1:1 or tetramorphic types. The kaolinite group represents 1:1 layer structures, because of its composition of one tetrahedral to one octahedral sheet. The montmorillonite group represents the 2:1 type, since its structure is built of two tetrahedral sheets to one octahedral sheet. The chlorite group is an example of a 2:2 type, while palygorskite and sepiolite belong to the 2:1:1 type.

Each clay mineral group can be subdivided again into two subgroups: (1) dioctahedral and (2) trioctahedral. If two of three of the octahedral positions are occupied by Al^{3+}, for example, it is dioctahedral. If all octahedral positions are occupied by Mg^{2+}, it is a trioctahedral subgroup.

In addition to the structural arrangement discussed above, stacking of the layers can also occur by different types of unit layers in a regular or irregular pattern. The latter yields the interstratified group of clay minerals. The structure of these minerals may vary widely, since two or more different types of unit layers may be stacked together, e.g., vermiculite with chlorite units, chlorite with smectite units, mica with smectite units, and kaolinite with smectite units.

Kaolinite Group (1:1 Layer Clays)

Kaolinite minerals are hydrated alumino-silicates with a general chemical composition $Al_2O_3:SiO_2:H_2O$ = 1:2:2, or $2SiO_2 \cdot Al_2O_3 \cdot 2H_2O$ per unit cell. As stated earlier, structurally they are 1:1 type phyllosilicates. The crystal is composed of aluminum octahedra sheets stacked above silica tetrahedron sheets (see Figure 4.20). The sheets extend continuously in the a and b directions and are stacked one above the other in the z or c direction. The unit cell is nonsymmetrical with a silica tetrahedra sheet on one side and an aluminum octahedra sheet on the other. Consequently, the basal plane of oxygen atoms in one crystal unit is opposite to the basal plane composed of OH ions of the next layer. The latter gives to the mineral two types of surfaces. The two sheets forming a unit layer are held together by oxygen atoms, which are mutually shared by the silicon and aluminum atoms in the respective sheets. The unit layers, in turn, are held together by hydrogen bonding, yielding an intermicellar space with fixed dimension. The basal (001) spacing of kaolinite is therefore 7.14 Å.

There is little isomorphous substitutions and the permanent charge per unit cell, if not zero, is very small. However, due to the presence of exposed hydroxyl groups, kaolinite has a variable, or pH-dependent, negative charge. As can be noticed from its structure, the position of the OH groups opens possibilities for dissociation of H^+, which is the reason for the development of the variable charges, especially the plane of hydroxyl groups on the exposed surface of the octahedral site. Another plane of hydroxyls is also present, but the latter is located as a subsurface plane of the octahedrons, covered by a network of oxygen ions. The possibility for dissociation of H^+ ions through such a network of oxygens is still unknown. The cation exchange capacity of kaolinite is, therefore, very small and may change with pH. Usually, it is in the range of 1 to 10 mEq per 100 g.

Because of the tightness of the structural bonds, kaolinite particles are not easily broken down. This is also the cause for low plasticity, shrinkage and swelling properties. Its restricted surface limits the adsorption capacity for cations. The specific surface area is approximately 7 to 30 m^2/g.

The presence of kaolinite can be identified by a (001) d spacing of 7.14 Å x-ray diffraction, and by a second-order 3.57 Å diffraction in oriented samples. These diffraction peaks will disappear after heating kaolinte to 500 to 550°C.

Members of the kaolinite group are kaolinite, dickite, nackrite, and halloysite. Except for halloysite, the other minerals are nonexpandable in water. Halloysite contains interlayer water as will be

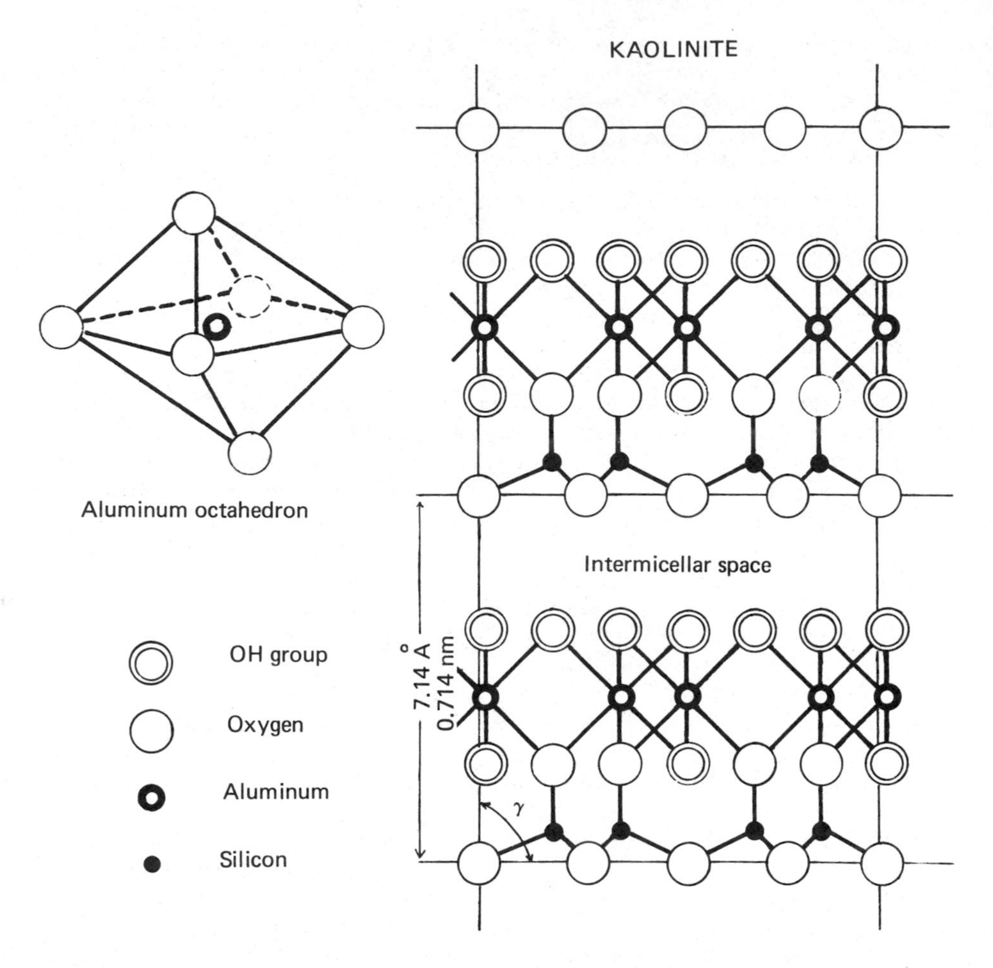

Figure 4.20 The structure of kaolinite, composed of silica-tetrahedron and aluminum-octahedron sheets looking down the (b) direction. Unit cell formula: $[Al_2(OH)_4(Si_2O_3)]_2$, a = 5.14 Å, b = 8.93 Å, c = 7.37 Å, α = 91.8°, β = 104.5°, γ = 90°.

discussed in the following pages. Upon heating, it is irreversibly dehydrated and the mineral is called metahalloysite. Of the mineral species listed above, kaolinite is perhaps the most widely distributed in soils. It is an important fraction of the clay of Ultisols and Oxisols, and is also detected as accessory minerals in Alfisols and Vertisols in the tropics.

Halloysite (1:1 Layer Clays)

This mineral has a general composition $Al_2O_3 \cdot 2SiO_2 \cdot 4H_2O$, and is similar in structure to kaolinite. The differences lie in the disordered stacking of layers, and in the presence of two or more interlayer water, as noted earlier for halloysite.

The water molecules are linked together in a hexagonal pattern. They are in turn bonded to the crystal layers by H bonding. Because of the presence of interlayer water, halloysite exhibits a basal (001) spacing of 10.1 Å, which upon heating can be reduced to 7.2 Å (see Figure 4.21). The dehydrated species is called metahalloysite, Halloysite is reported to convert rapidly into metahalloysite at 50°C, but the basal d spacing will collapse only after heating at 400°C. Although heating reduces the d spacing, it does not affect the random stacking of layers.

Halloysite is in general tubular in form, as shown by electron microscopy (see Figure 4.22). This is in contrast to kaolinite, which is hexagonal in shape. However, recently sheetlike halloysite, called *tabular* halloysite, has been detected in some of the soils in Texas. The tubular crystal form is considered to be rolled up sheets. Due to the presence of interlayer water, the normal z-axis bonding of O–OH groups is prevented, causing a distortion of the crystal structure with the consequent curling of layers.

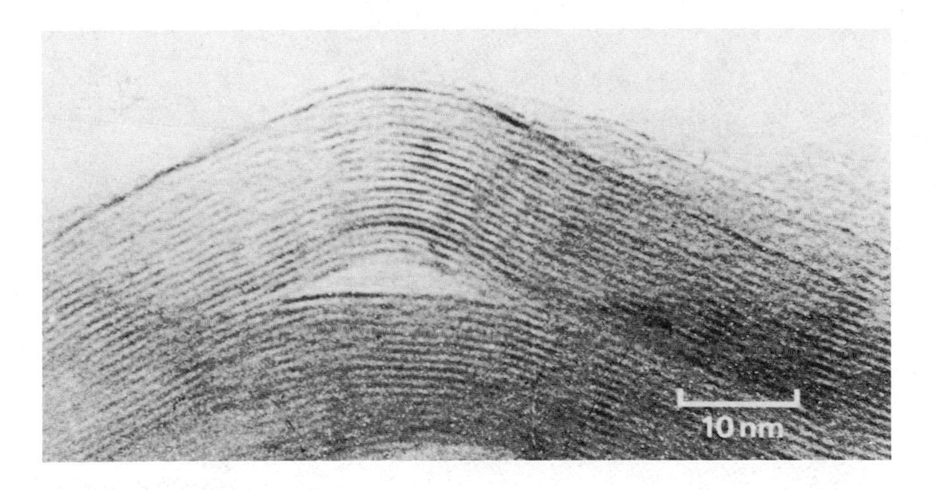

Figure 4.21 Transmission electron microscopy of halloysite, showing lines, corresponding to basal unit layers, repeating at distances of approximately 7.2 Å (0.72 nm). [From Sudo and Yotsumoto (1977), Figure 3(b). Courtesy of Clays and Clay Minerals.]

Figure 4.22 Transmission electron micrographs of kaolinite (left) and halloysite (right) minerals. [From Egawa and Watanabe (1964).]

The x-ray diffraction pattern of halloysite, dried at 105°C, is almost similar to kaolinite. However, the basal (001) diffraction peak of halloysite is usually broad or less sharp, due to the disordered stacking of layers. Partially dehydrated halloysite may exhibit x-ray diffraction patterns between the two end members stated above (between 10.1 and 7.2 Å).

Formation and stability of halloysite in soils appear to be influenced by soil moisture. A moist condition is considered to be required for the development of this mineral. Indications are present that halloysite can be considered a precursor of kaolinite, since formation of the mineral follows the weathering sequence: igneous rock → montmorillonite → halloysite → methahalloysite → kaolinite.

Montmorillonite (Expanding 2:1 Layer Clays)

Minerals in this group are sometimes called *smectite*, and have a variable composition. However, the formula is often expressed as $Al_2O_3 \cdot 4SiO_2 \cdot H_2O + xH_2O$. The name montmorillonite is reserved for the hydrated aluminosilicate species with little substitution. Many clay deposits in the USA contain large amounts of montmorillonite. This type of clay is frequently called *bentonite*, and commercial grade montmorillonite is also often referred to as bentonite.

A wide range of minerals exists within the montmorillonite group, and the principle end members in the dioctahedral subgroup are beidellite and nontronite. Montmorillonite has Mg and ferric ions in octahedral positions, whereas beidellite ideally contains no Mg or Fe in the octahedral sheet. Beidellite is characterized by a high Al content. The silicate layer charge is derived entirely by substitution of Al^{3+} for Si^{4+}. Nontronite is like beidellite, but with all the Al^{3+} replaced by Fe^{3+}. In the trioctahedral subgroup, only two end members are recognized, e.g., hectorite and saponite.

Two types of structure have been proposed for montmorillonite, e.g., the structure according to (1) Hofmann and Endell and (2) Edelman and Favajee. Both hypotheses show similarity in the fact that the unit cell structure is considered symmetrical, as opposed to that of kaolinite. One aluminum octahedral sheet is sandwiched between two silica tetrahedra sheets. The crystal layers are reported to be stacked together in random fashion, while some of the minerals may even be fibrous, such as hectorite. The bonds holding the layers together are relatively weak, developing intermicellar spaces that will expand with increasing moisture content (see figure). However, the difference between the structure of Hofmann and Endell and that of Edelman and Favajee is in the arrangement of the silica tetrahedra network, as shown in Figure 4.23. Edelman and Favajee are of the opinion that an alternative arrangement of silica tetrahedra exists with a Si-O-Si bond angle = 180°, with basal planes composed of OH groups bonded by the silica in the tetrahedrons.

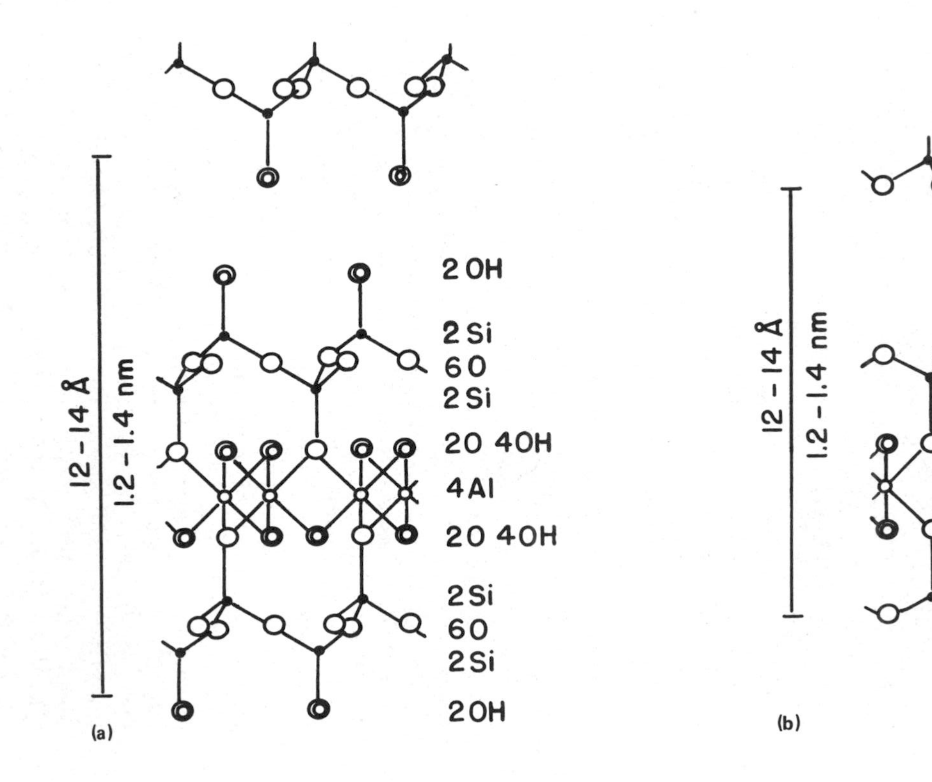

Figure 4.23 (a) Structural model according to Edelman and Favajee, and (b) structural model according to Hofmann and Endell. [From M. Wisaksono Wirjodihardjo and K.H. Tan (1964). *Ilmu Tanah.* Djilid II. Publishers Pradnjaparamita II, Jakarta. Indonesia.]

The negative charge of montmorillonite arises mainly from isomorphous substitution. Only a small variable charge is present, since all the disposable hydroxyl groups are located in subsurface planes covered by a network of oxygen atoms. Van Olphen (1977) mentioned a charge equivalent to a cation exchange capacity of 70 mEq per 100 g for a typical montmorillonite. The specific surface area is approximately 700 to 800 m^2/g, and because of this large specific surface area, which is exposed on dispersion in water, montmorillonite exhibits strong plasticity and stickiness when wet.

The minerals are generally very fine grained, while the component layers are not bonded strongly, as noted earlier. In contact with water, the mineral exhibits interlayer swelling, causing the volume of the clay to double. Indications are available that the basal spacing of montmorillonite increases uniformly with adsorption of water. Several authors noted that the increase in basal spacing can occur stepwise, suggesting formation of hydration shells around interlayer cations. The high swell-shrink potential is the reason that the mineral can admit and fix metal ions and organic compounds. The adsorption of organic compounds leads to formation of organo-mineral complexes. Organic ions are believed to be able to replace inorganic cations in the interlayer position. Monolayers, sometimes double layers, of organic molecules are adsorbed depending on the size of the cations and the charge deficit of the layers.

Adsorption of organic compounds, such as glycerol and ethylene glycol, is diagnostic in identification of montmorillonite by x-ray diffraction analysis. Ovendried (105°C) montmorillonite is usually characterized by a basal (001) spacing diffraction peak at 10 Å. In air-dry conditions, the mineral has some interlayer water, and the characteristic spacing is approximately 12.4 to 14 Å (see Figure 4.24). After intercalation with ethylene glycol or glycerol, the basal (001) d spacing expands to 17.0 Å. Reports are present in the literature that suggest that the spacings can be expanded indefinitely. However, in this case the mineral exists only as platelets or as unit cells, while the intermicellar spaces cease to exist.

Among the several species of clays in soils, the mineral montmorillonite is perhaps the most widely distributed member. Most soil montmorillonites are dioctahedral. They are characteristic constituents of clays of Vertisols, Mollisols, and Alfisols and are also found in some Entisols. The high plasticity and swell-shrink potential of the mineral make these soils plastic when wet and hard when dry, while wide cracks will form as soils dry out. The dry soil is difficult to till.

Illites (Nonexpanding 2:1 Layer Clays)

These minerals are micaceous types of clay. However, in contrast to the true micas, which are of primary formation, illites are of secondary origin. They are also known under the names hydrous mica or soil

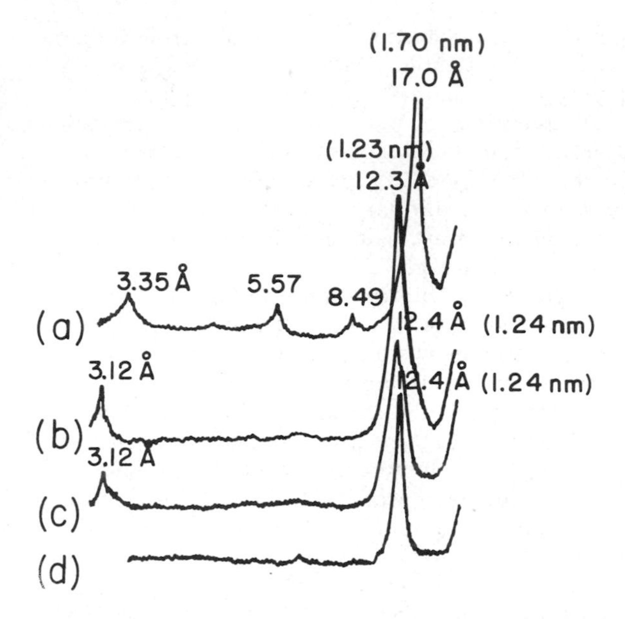

Figure 4.24 X-ray diffraction analysis of montmorillonite: (a) after solvation, (b) through (d) air dry. [From Tan and McCreery (1975).]

mica. The term *illite* is suggested for the fine-grained minerals in this group, while the coarser particles are frequently called *hydrous mica*.

A number of authors object to classifying illite as clay. They indicate that, strictly speaking, illite is a clay-size mica and, therefore cannot be regarded as a clay mineral (Fanning and Keramidas, 1977; Theng, 1974). However, illitic clay mineralogy is recognized by the U.S. soil taxonomy (USDA, 1975). Van Olphen (1977) considers mica, especially muscovite, the prototype of illite. Its close relation with mica was the reason for naming this mineral hydrous mica or soil mica.

The mineral has an almost similar chemical composition as muscovite, but contains more SiO_2 and less K. Alteration, associated with weathering processes of muscovite and subsequent formation into clays, are the reasons for the observed differences. Several authors are of the opinion that a continuous series of illite species exists between muscovite and montmorillonite as the end members. Mixed layering of illite-montmorillonite often occurs.

$$H_2KAl_3Si_3O_{12} \leftarrow - - - - - - - - - - - - \rightarrow Al_2O_3 \cdot 4SiO_2 \cdot H_2O + xH_2O$$

Muscovite illite series Montmorillonite

Since illite contains interlayer potassium, the unit layers are bonded stronger together than montmorillonite. Therefore, the intermicellar spaces of illite do not expand upon addition of water. Because of the electrostatic bonds exerted by K^+ ions linking the unit layers together, the basal (001) spacing is 10 Å. The cation exchange capacity is in the order of 30 mEq per 100 g, and plasticity, swelling, and shrinking are less intense in illite than in montmorillonite. Fine-grain illites have been found concentrated in the coarse clay fraction (2 to 0.2 μm) of soils. As indicated earlier, the physical properties are closer to kaolinite than to montmorillonite. However, the ease of parallel alignment of the particles and their presence in the coarse clay fraction are considered to have a detrimental effect on soil stability.

Identification of illite can easily be done by x-ray diffraction analysis. These minerals are characterized by a basal (001) spacing of 10.0 Å. The latter peak does not shift or collapse after heating of the mineral at 500°C or after solvation with glycerol, glycol, or ethylene glycol. In many instances, elemental analysis of K concentration has also been used for the detection of illites in soil clays. Although the theoretical concentration is approximately 9 to 10% (K), a potassium content of 5 to 8% is frequently found for illitic clays, with a value of 7% (K) as the diagnostic percentage for illite.

Illite has been found an important constituent of clays in Mollisols, Alfisols, Spodosols, Aridisols, Inceptisols, and Entisols. In soils affected by high precipitation, the mineral tends to be altered into montmorillonite, whereas under the influence of warmer climates or higher temperatures, the structure of illite is reported to become more disordered, and kaolinite is formed.

Vermiculites

This group of minerals also forms mica-like flakes as illites. It is also a mica alteration product. However, the mineral called *hydrobiotite*, formed from weathering of biotite, is not vermiculite, but belongs to the illite group.

Vermiculite can be divided into two categories: true vermiculite and clay vermiculite. True vermiculite is not considered a clay mineral, but a rock-forming mineral (Douglas, 1977; Walker, 1975). The name is derived from *vermiculare* or *vermicularis* (Latin, wormlike, or to breed worm), since upon heating the mineral becomes elongated, twisted, and curved. After heating, it usually expands to 20 to 30 times its original size. Commercial vermiculite is often interlayered biotite and vermiculite. The clay-size vermiculite found in soils is considered "clay vermiculite" or "soil vermiculite." Its existence in the clay fraction of soils was first demonstrated in 1947 in the soils of Scotland, but Walker (1975) indicated that it has not been isolated yet as monomineral particles. The detection in soils is based on its x-ray diffraction peak at 14 Å. The latter is the reason for sometimes calling this mineral 14 Å mineral.

Clay vermiculite is a magnesium aluminum silicate, with Mg occupying the octahedral positions between two silica tetrahedra sheets. Some iron may also be present. The chemical formula can be generalized as

$$22MgO \cdot 5Al_2O_3 \cdot Fe_2O_3 \cdot 22SiO_2 \cdot 40H_2O$$

or

$$Mg_3Si_4O_{10}(OH)_2 \; x \; H_2O$$

The structure (see Figure 4.25) shows close similarities with that of chlorite, with the difference that, instead of brucite, water molecules of about 5 Å layers are occupying the intermicellar spaces. In many cases, interlayering with hydroxy-Al also occurs. In the tetrahedral layer considerable substitution of Al for Si takes place. This is the reason for the high negative charge present in vermiculite.

Vermiculite is one of the clay minerals with the largest CEC among the inorganic colloids. The cation exchange capacity is approximately 150 mEq per 100 g, and exceeds that of montmorillonite. According to Douglas (1977), the CEC of diocathedral vermiculite is 1.05 times the CEC of trioctahedral vermiculite, and values of CEC between 144 and

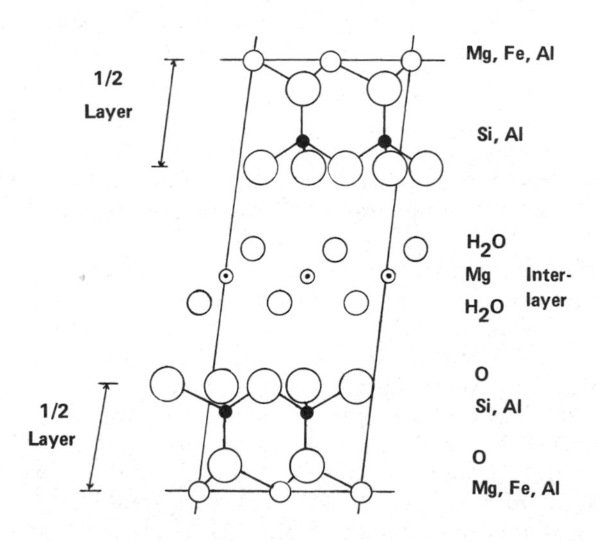

Figure 4.25 The structure of vermiculite, showing the brucite interlayer.

207 mEq per 100 g have been reported. The presence of hydroxy-Al interlayers usually reduces the CEC of the mineral. Soils of the southern region of the United States, often characterized by low CEC, exhibit higher CEC values by small admixtures of vermiculite in their clay fraction.

Most soil vermiculites are probably dioctahedral. The mineral is reported to have wedge zones with high selectivity for fixation of K^+, NH_4^+, and other cations. The high potassium and ammonium fixation values in many soils are attributed more to the presence of vermiculite than to montmorillonite or illitic type of clays.

Identification of clay vermiculites is done mostly by x-ray diffraction analysis and differential thermal analysis (DTA). Using oriented samples the basal (001) x-ray diffraction peak is 14 Å. This peak will not shift or collapse upon solvation, but after heating to 700°C, the basal d spacing usually collapses to 11.8 or 9.3 Å. The common occurrence of vermiculite as mixed layers with montmorillonite, chlorite, illite, and biotite, yields many difficulties in the positive identification of vermiculite. Many clay vermiculites may have been identified in the past as montmorillonites. In some instances treatment of vermiculite with KCl solutions can produce a mineral with a mica structure.

Vermiculite usually occurs as accessory minerals in the clay fractions of Ultisols, Mollisols, and Aridisols. It is formed more in well-drained soils, in contrast to montmorillonite which requires a gley condition for formation.

Chlorites (2:2 Layer Clays)

Chlorites are hydrated magnesium and aluminum silicates, which are related to mica minerals in appearance. The name comes from the green color of many chlorite specimens. Structurally chlorite is related to talc, or 2:1 layer clays and shows close relationship with vermiculite. However, recently a number of authors prefer to use the term *2:2 layer* for chlorite. Octahedral sheets, composed of Mg $(OH)_2$ are sandwiched between the two silica tetrahedra sheets. The $Mg(OH)_2$ sheet was formerly called the *brucite* sheet. The intermicellar spaces are also occupied by brucite sheets, hence the term 2:2 layer clays.

The mineral is variable in composition, but the general composition is reported as follows:

$$(Mg, Fe, Al)_6(Si, Al)_4O_{10}(OH)_8$$

Isomorphous substitutions occur in both the tetrahedral and octahedral layers. The silicon may be replaced by Al, while iron and/or Al may replace Mg in octahedral positions. The degree of substitution is expressed by Foster (1962) as Fe^{2+}/R^{2+} ratios, and on this basis three general groups of chlorites are recognized (see Figure 4.26):

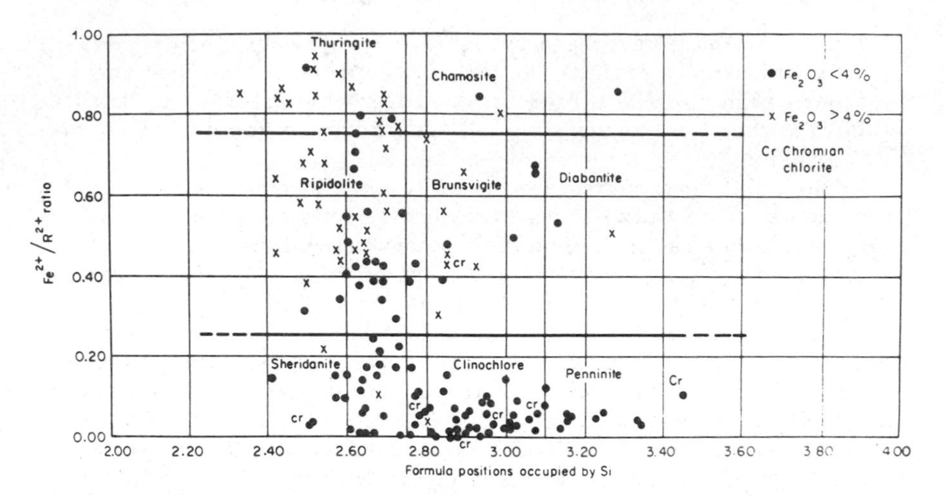

Figure 4.26 Classification of chlorites on the basis of degree of isomorphous substitution as expressed in Fe^{2+}/R^{2+} ratios. [From Foster (1962).]

(1) Fe chlorites, containing relatively high amounts of iron, (2) intermediates, and (3) Mg chlorites, which contain smaller amounts of iron. Other types of chloritic minerals have also been recognized, e.g., swelling chlorites or corrensites. The latter swells when wet. They are supposed to be more mixed-layer minerals composed of chlorite-montmorillonite and vermiculite than normal chlorite.

The replacement of Mg by Al occurring in the brucite sheets accounts for the development of a positive charge. This positive charge practically neutralizes the negative charge of the "mica" layer. Therefore, chlorite has only a very small charge, and consequently a small CEC. The hydroxy interlayers are sites for anion retention. Phosphorus is reported to be fixed by interlayer hydroxides of chlorites. On the other hand, the presence of these interlayers reduces fixation of K and/or NH_4^+ ions.

Depending on the species, the characteristic d (001) spacing of chlorite is 14.0 Å, as determined by x-ray diffraction analysis using oriented specimens. This peak does not shift or collapse by treatments of the sample with glycerol or ethylene glycol or by heating to 500°C. Swelling chlorite may have a basal spacing of 28 Å, which increases to 32 Å by solvation.

Abundance and frequency of occurrence of chlorite are considered low (Barnhisel, 1977). Chlorite is usually detected as accessory minerals in clays of Alfisols, Mollisols, and Aridisols. Most of the chlorite minerals are trioctahedral, but recently dioctahedral chlorite has been detected in Virginia from soils derived from muscovite-schist and in soils from British Columbia.

Mixed-Layer Clays

Soil clays exist in nature as a mixture of several species. Some of the different types of clay minerals may be stacked together as a packet. This is called interstratification, and such clays are referred to as *interstratified clays* or mixed-layer clays. Interstratified clays cannot be separated by physical means as is the case with an ordinary mixture of clays.

Interstratification can occur in a (1) regular fashion or (2) random fashion. It may also be the result of a segregation process within a crystal in zones within another mineral. Another process of formation of mixed-layer clays is precipitation, formation, or crystal growth in interlayers. For example, gibbsite sheets may develop from precipitation and crystallization in intermicellar spaces, due to replacement of exchangeable cations and change in chemical environment.

The identification of regular mixed-layer clays follows the same principles as used with monomineral clays. X-ray analysis also produced a regular sequence of x-ray diffractions with regular mixed clays. They are usually identified by an integral sequence of the basal (001) diffraction peaks, which corresponds to the sum of the thickness of the component layers. For example, two layers of 14 Å vermiculitic clay will yield a basal d spacing of $2 \times 14 = 28$ Å. Corrensite is considered a regularly interstratified chlorite-montmorillonite clay. In this case, the basal (001) diffraction peak is either $14.0 + 17.0 = 31.0$ Å (for an expanded montmorillonite component), or $14.0 + 12.4 = 26.4$ Å (for an air-dry montmorillonite component), assuming that the structure has not been changed.

The random mixed layer clays are more difficult to identify, and do not exhibit integral series of basal (001) diffraction patterns. The sequence of diffraction is relatively short compared to regular mixtures. Pretreatments, such as solvation, potassium saturation, and heating, may then be required to solve the problem.

On the other hand, physically mixed clays can be readily identified, since the diffraction of the basal (001) spacings of the major planes will all appear in the x-ray analysis. Each of the diffraction peaks can be identified in the usual manner.

Interstratified clays have been detected in a large variety of soils in temperate, cold, and tropical regions. MacEwan and Riuz-Amil (1975) are of the opinion that mixed layering is less common in tropical conditions. In the soils of the humid temperate regions, interstratification occurs often in the sequence montmorillonite-chlorite-mica, or mica-illite. However, montmorillonite-kaolinite, or vermiculite-kaolinite has also been observed, especially in the subtropical regions of the United States. In the humid tropics, interstratification has been noted in the sequence montmorillonite-halloysite-kaolinite. In the United States, chlorite and vermiculite are the most common interstratified clays in Alfisols and some Ultisols.

Silica Minerals

Silica minerals are minerals composed entirely of silica. They occur extensively in nature and are frequently an important constituent of the clay fractions of soils. However, the coarse silica particles are found mostly in the silt and sand fractions.

Structurally these minerals do not belong to the phyllosilicates characterized by sheet structures, but they are distinguished as minerals with framework structures or tectosilicates. The four oxygen atoms of the silica tetrahedron are linked directly to neighboring silicon atoms, yielding a fourfold coordination, which is electrically balanced. The formula of these minerals is generalized as $n(SiO_2)$. Three types of minerals are distinguished in the category of silica minerals, e.g., quartz, tridymite, and crystobalite. Depending on the temperature, each of them can exist in an α and β. The α form is the low-temperature variety, whereas the β modification is the high-temperature form. The transformation, called *conversion*, is normally instantaneous and reversible, and accompanied by structural changes (see Figure 4.27).

α Quartz (trigonal) α tridymite (hexagonal) α crystobalite (tetragonal)

$\upharpoonleft\!\downharpoonright$ 573°C $\upharpoonleft\!\downharpoonright$ 117°C $\upharpoonleft\!\downharpoonright$ 220-280°C

β Quartz (hexagonal) $\rightleftharpoons$ β tridymite (hexagonal) $\rightleftharpoons$ β crystobalite (cubic)

870°C   1470°C

The silica minerals are generally considered inert, or chemically inactive, material. They have only a slight effect on physicochemical properties of soils, and because of their low chemical activity, they are perhaps of importance only as a diluent to the more reactive clay and humic material. The surface area is very small, and amounts only to 2 to 3.0 m^2/g, depending on the shape of the particles. Soils with clay fractions dominated by silica minerals, are usually nonplastic and have a small shrink-swell capacity, as well as a small water holding capacity. The surface charge is also very small, if not negible, while the correspondingly small cation and anion exchange are attributed more to the Si–O broken bonds and Si–OH groups on particle edges.

Silica minerals are insoluble at low pH. Their solubility does not increase if pH is increased, e.g., from pH 3 to pH 9 (Krauskopf, 1956). Only above soil pH 9.0, will silica dissolve according to the following reaction:

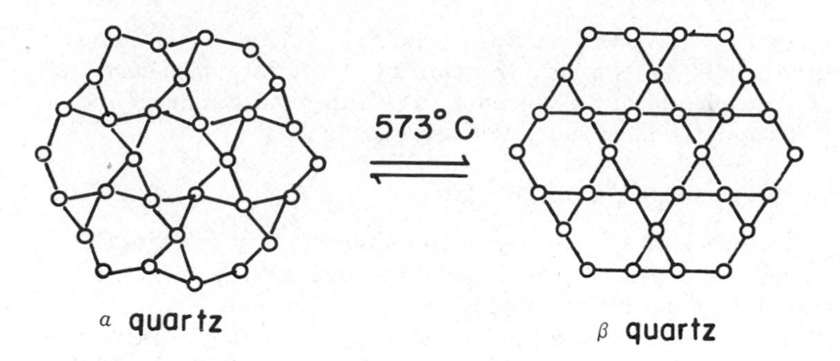

Figure 4.27 Structural changes associated with conversion of α to β quartz.

$$Si(OH)_4 + OH^- \rightleftharpoons Si(OH)_3O^- + H_2O$$

In general, solubility of silica minerals is related to the packing density of the silica tetrahedra. Of the three types of minerals stated, solubility increases in the following order:

Quartz < crystobalite < opal < amorphous silica

The sequence suggests that quartz is the least soluble, whereas amorphous silica is the most soluble form. Opal is mostly of plant origin and is also called *biogenic silica.*

Identification of silica minerals is usually done by x-ray diffraction and DTA. With x-ray analysis quartz yields a d (100) spacing of 4.26 Å and a d (101) spacing of 3.34 Å. Frequently the d (100) diffraction peak at 4.26 Å is very weak in intensity, leaving the usually strong 3.34 Å x-ray diffraction peak for use in diagnosis of quartz. With DTA, quartz exhibits a small but sharp endothermic peak at 573°C. Since the temperature at which this endothermic peak occurs is sharp at 573°C, quartz is often used as a stable reference material for calibration of the DTA instrument. The identification of crystobalite is more difficult than quartz, since crystobalite gives a series of d spacings by x-ray analysis, e.g., 4.04, 3.14, and 2.84 Å, which overlap the d spacings of orthoclase minerals. Generally, a strong 4.04 Å peak accompanied by a relatively weak 3.14 Å peak suggests the presence of crystobalite.

The silica minerals occur in a wide variety of soils. Their contents appear to be related to parent material and to degree of weathering. Quartz is usually an important mineral in Spodosols, because of the formation of these soils from highly siliceous material. Inceptisols may be rich in quartz, but this may also be a reflection of the parent material. In moderately weathered Alfisols, Ultisols and Mollisols, quartz may accumulate in the eluvial horizons. On the other hand, quartz may be absent in the clay fraction of the highly weathered Oxisols. Crystobalite is often volcanic in origin, and its presence is considered of importance in many volcanic ash soils.

Iron and Aluminum Hydrous Oxide Clays

This group of clays is currently becoming increasingly important. They do not belong to the phyllosilicates but are oxides of iron and aluminum containing associated water.

Two major forms of crystalline monohydrates of ferric oxide are known, e.g., goethite and lepidocrocite, and two crystalline anhydrous ferric oxides have also been found in soils. They are hematite and maghemite. The composition of these and other hydrous oxide minerals are listed in the following table.

Goethite	α-FeOOH	Diaspore	α-AlOOH
Lepidocrocite	γ-FeOOH	Boehmite	γ-AlOOH
Hematite	α-Fe_2O_3	Gibbsite	$Al(OH)_3$
Maghemite	γ-Fe_2O_3		
Ferrihydrite	$Fe_5HO_8 \cdot 4H_2O$ or $Fe_5(O_4H_3)_3$		

Limonite has also been frequently mentioned in the literature as an important rusty iron oxide mineral. However, it is currently no longer considered a soil mineral. Recently a newly discovered hydrous iron oxide mineral, ferrihydrite, is gaining in importance. According to Schwertmann and Taylor (1977), in the past it has been erroneously called "amorphous ferric hydroxide." It was found as a major component of iron ochre sediments in drainage ditches, and is suspected to also occur widely in soils. Perhaps the B_{ir} horizons of Spodosols contain clay fractions with ferrihydrite.

Structurally goethite is formed by close-packed oxygen atoms in a hexagonal pattern (see Figure 4.28). On the other hand, lepidocrocite has a more complicated structural pattern. Isomorphic substitution of Al and/or Mn for some of the Fe frequently occurs.

The most common aluminum hydrous oxide in soils is gibbsite. Gibbsite is sometimes also called *hydrargillite*. Less common aluminum hydrous oxide species are bayerite, boehmite, and diaspore.

The composition of gibbsite is usually formulated as $Al(OH)_3$, with a structure made up of layers composed of two close-packed hydroxyl sheets with aluminum located in a sixfold coordination (see Figure 4.29). The hydroxyl groups are arranged in a slightly polar position in the structure. The Al^{3+} ions occupy two-thirds of the

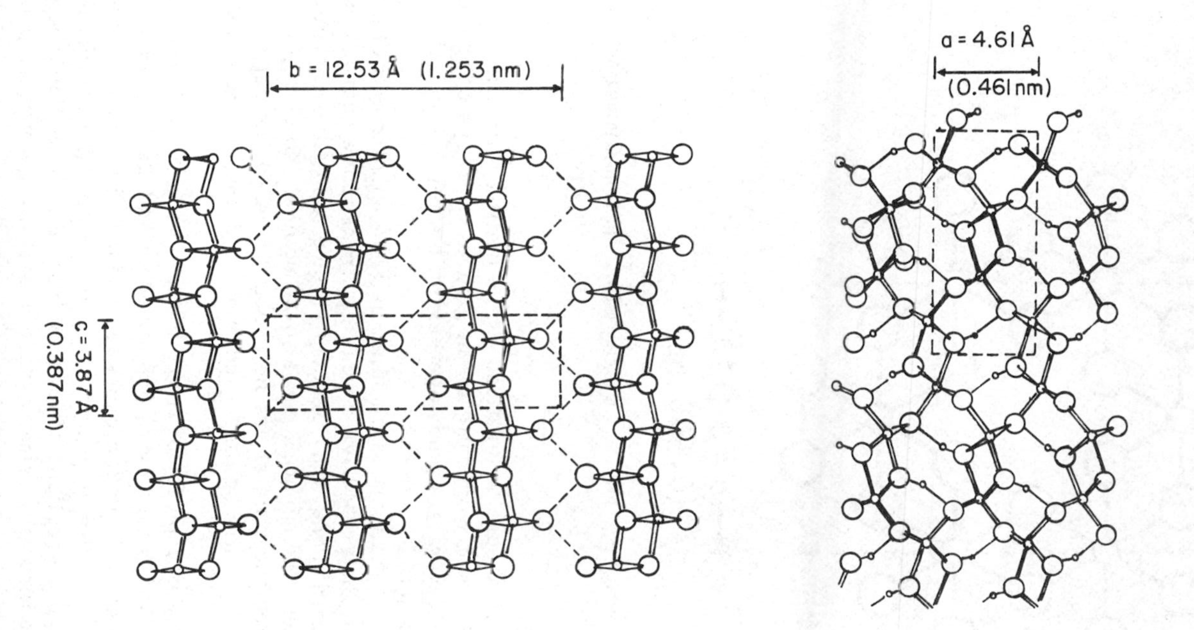

Figure 4.28 Structure of goethite along the c axis. [From Greenland and Hayes (1978), p. 139. Reprinted by permission of John Wiley & Sons, Ltd.]

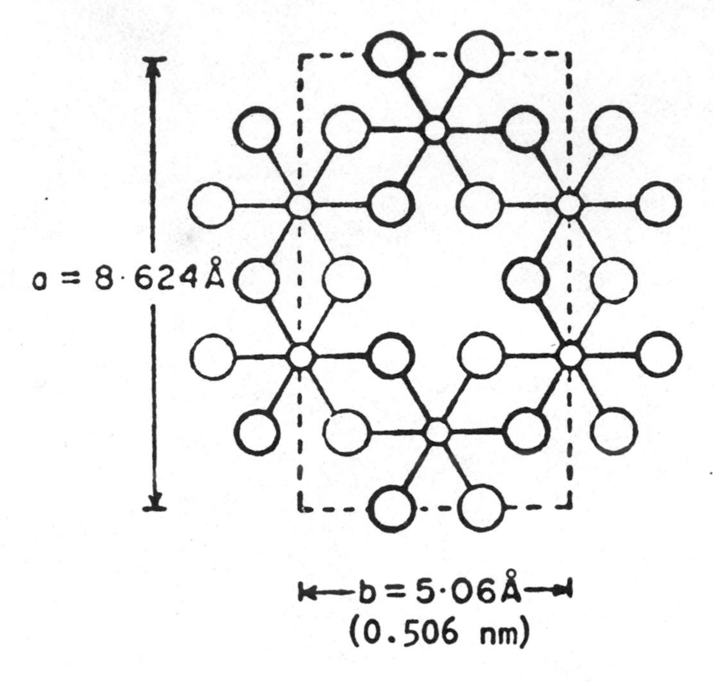

Figure 4.29 Schematic structure of gibbsite. [From Grimshaw (1971).]

possible vacant octahedral interstices. The hydroxyl groups of one layer are almost directly opposite to the hydroxyl groups of the adjacent layer. The layers are held together by hydrogen bonds between opposite OH groups.

The iron and aluminum oxide minerals are amphoteric in character. In acid condition they may have a weak electronegative charge, and in alkaline soil condition they may develop an electropositive charge. At certain pH values, the minerals can also be neutral (no charge). The pH value at which the mineral has no charge is called the *zero point charge* (ZPC). This will be discussed more in detail in a later section of this book.

The adsorption capacity of iron minerals ranges from 30 to 300 μmol per gram, and according to Schwertmann and Taylor (1977) this compares favorably with cation exchange capacities of silicate minerals. The latter authors distinguish between "nonspecific" and "specific" adsorption or "chemisorption" of ions by iron minerals. Nonspecific adsorption is defined as an electrostatic adsorption, whereas specific adsorption is related to a covalent type of bonding of ions. Specific adsorption occurs with phosphate ions and heavy metal cations, such as Cu, Zn, Mn, and Pb, leading to a reaction called *retention* or *fixation*. It has been reported that adsorption of HPO_4^{-2} increases the negative charge of the mineral, and consequently the CEC will also increase.

The iron oxide minerals have also been known to influence the physical properties of soils. Indications are available that iron oxides are adsorbed on kaolinite surfaces, inducing a cementation effect, leading to the consequent development of strong aggregation of soils particles and to concretion and crust formation (Baver, 1963).

The identification of the iron and aluminum oxide minerals is done by x-ray diffraction and DTA. X-ray analysis yields strong diffraction peaks at 4.18 Å and at 4.82 Å for goethite and gibbsite, respectively. Differential thermal analysis reveals a strong endothermic peak at approximately 290 to 350°C for goethite, gibbsite, and the other minerals. The problem is usually solved by differential dissolution with NaOH of the clay mixture. This treatment will dissolve the gibbsite components, and an endothermic peak at the temperature range as indicated above in DTA curves of the residue is then indicative for the presence of goethite and/or other iron minerals. Scanning electron microscopy shows gibbsite to be rhombohedral in shape (see Figure 4.30).

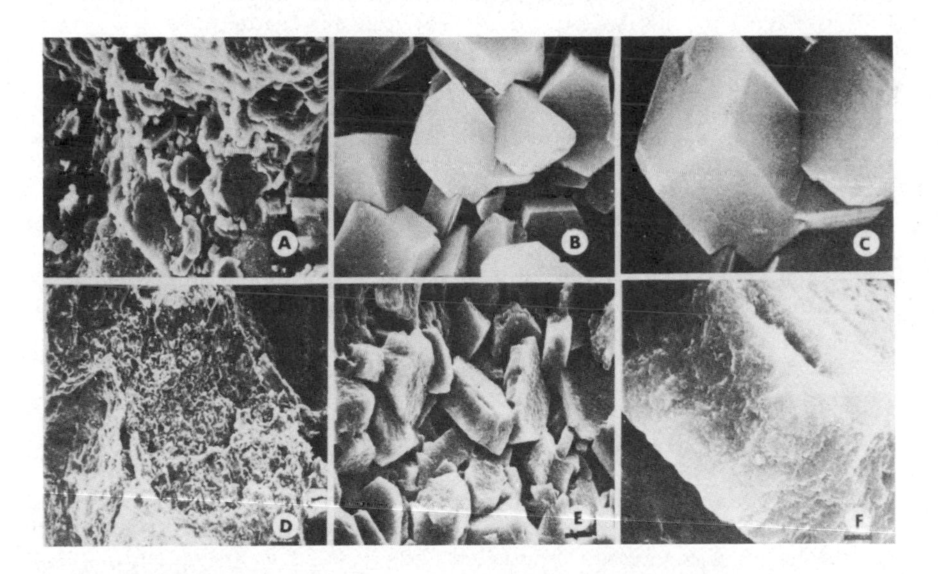

Figure 4.30 Scanning electron microscopy of gibbsite crystals. A. Magnification is × 500, B. × 5,000, C. × 10,000, D. × 100, E. × 1,000, and F. × 5,000 [From H. Eswaran et al. (1977).]

Goethite, hematite, and gibbsite are probably the most frequently found forms of iron and aluminum oxides in soils. They may occur in considerable amounts in the clay fraction of especially tropical and subtropical soils. Many authors consider their presence to be an indication of the effect of drastic weathering processes. The red and yellow colors of highly weathered soils are attributed to these minerals.

Goethite is the most important iron oxide mineral in many soils and is responsible for the reddish-brown colors. Hematite, on the other hand, is of lesser importance, but it may occur in tropical and subtropical soils. This mineral is the reason for the red colors of many tropical soils. Gibbsite is a major mineral in highly weathered Ultisols and Oxisols of the humid tropical and subtropical regions. Bauxite deposits in tropical regions contain mostly gibbsite.

Amorphous Clays, Allophane, and Imogolite

With the recent progress in clay mineralogy, it is currently known that many soils also contain amorphous clays. These clays are noncrystalline in nature and include a wide variety of materials, e.g., silica gel, sesquioxide gels, silicates, and phosphates. They are amorphous to x-ray diffraction analysis. This means that they exhibit a featureless x-ray pattern. Udo Schwertmann (personal communications) is of the opinion that the term *amorphous* is used erroneously in soil science. It is perhaps the methods of analysis which are inadequate to detect crystallinity in the so-called amorphous clays, since a majority of these clays occur as very fine, micro, crystal forms. Wada (1977) prefers the use of the term *noncrystalline* over the term *amorphous*.

The most important type of clay in this group is perhaps allophane. It is found especially in volcanic ash soils. The name allophane was first introduced by Stromeyer and Hausmann in 1861 for hydrous aluminosilicates occurring in nature. Since then the name allophane has found general acceptance for a wide variety of clay material amorphous to x-ray diffraction analysis (Ross and Kerr, 1934). Allophane was formerly classified as kaolin clay since it has a sheet structure similar to that of kaolinite. The chemical composition is characterized by a molecular Al:Si ratio of 1:1 or 1:2. The formula is proposed to be

$$SiO_2 \cdot Al_2O_3 \cdot 2H_2O \quad \text{or} \quad Al_2O_3 \cdot 2SiO_2 \cdot H_2O$$

Another important type of clay in this group is imogolite. This clay mineral has been reported for the first time in 1962. It was found in weathered volcanic ash or pumice beds, called *imogo* (Yoshinaga and Aomine, 1962). Since then it has been detected in many volcanic ash soils in Japan, South America, and in the islands in the Pacific.

The composition formula of imogolite is assumed to be

$SiO_2 \cdot Al_2O_3 \cdot 2.5H_2O$

In many respects imogolite has chemical characteristics similar to allophane. Several authors believe that allophane is a precursor of imogolite. However, in contrast to allophane, imogolite has a better defined crystal shape. Electron microscopy shows evidence of the presence of hairlike or spaghettilike crystal forms. (See Figure 4.31.) The term *paracrystalline* has been suggested for the structure of imogolite.

The presence of allophane gives to the soil unique properties. Allophane has a large variable charge. It also behaves amphoterically and is reported to fix considerable amounts of phosphates. The CEC is approximately between 20 and 50 mEq per 100 g, whereas the AEC ranges from 5 to 30 mEq per 100 g. Imogolite has a larger CEC value than allophane. Wada (1977) estimated the CEC of imogolite to be 135 mEq per 100 g clay. As is the case with the iron and aluminum oxide minerals, anion adsorption by allophane is also divided into "nonspecific" and "specific" adsorption. Nonspecific adsorption also refers to electrostatic adsorption, whereas specific adsorption is the adsorption of ions due to covalent bonding in the coordination shells of the Al (or Fe) atoms. The amount of ions adsorbed nonspecifically increases with lower pH. Ions adsorbed by the specific process are considered to be fixed, or in other words, they can be replaced only with difficulty by other ions. This fixation process is of special importance in phosphate fixation.

The presence of allophane has also an important effect on a number of soil physical properties. Soils high in allophanic clays are characterized by low bulk density values, high plasticity, although they are nonsticky when wet. The waterholding capacity appears to be increased substantially by allophane.

It is assumed that allophane and imogolite will also undergo interaction processes with soil organic compounds, such as humic and fulvic acids. The latter reaction is called *complex formation* or *chelation*. Most soils containing allophane are known to have black A horizons, extremely high in soil organic matter content. These soils, in the past, were called *andosols* or *ando soils* (Japanese, ando-black).

Identification of these minerals is mostly done by DTA, since x-ray analysis yields featureless diffraction curves. Imogolite exhibits broad x-ray diffraction peaks at 12 Å, 7.8 to 8.0 Å, and at 5.5 Å in oriented speciments (Yoshinaga and Aomine, 1962), but the positive identification of imogolite is done by means of electron microscopy. As shown earlier, imogolite has a hairlike or spaghettilike shape in electron micrographs.

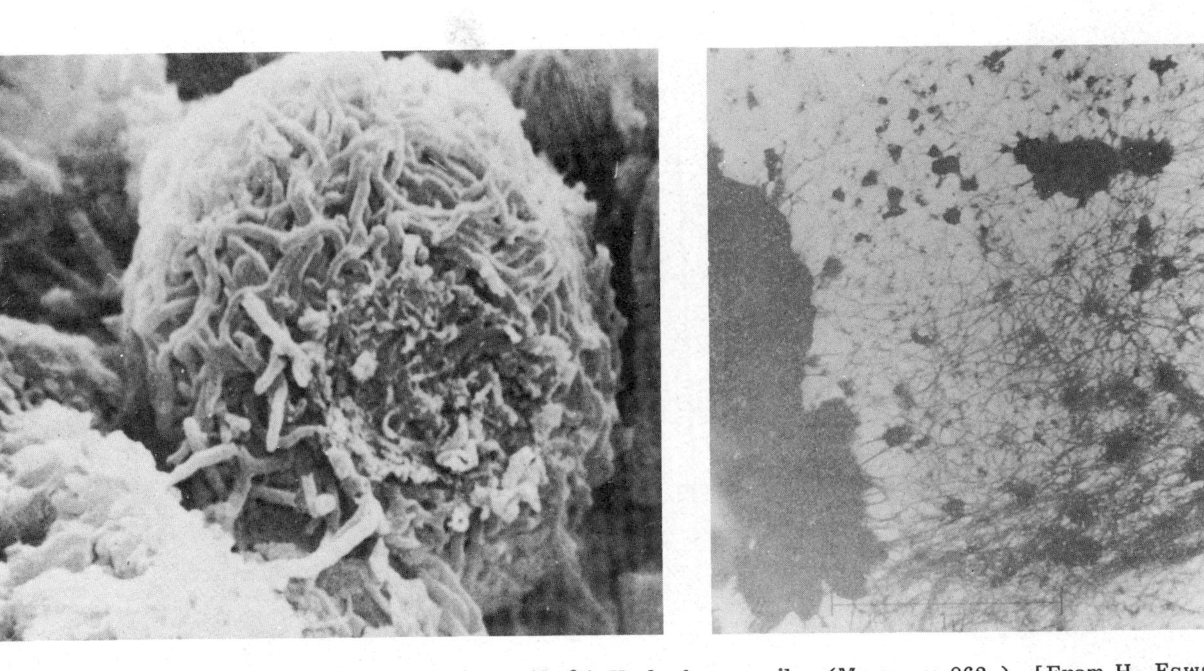

Figure 4.31 Electron microscopy of imogolite. (Left) Kodonbaru soil. (Magn. × 862.) [From H. Eswaran (1972). Clay Minerals 9:281-285] (Right) Volcanic ash soils in Chile. [From Besoain (1968).]

The common method for the detection of allophane is differential thermal analysis. The DTA curve of allophane is generally characterized by a large and sharp endothermic peak between 50 and 200°C (323 and 473 K), attributed to lost of adsorbed water, and a sharp exothermic peak at 900 to 1000°C (1173 to 1273 K), due to formation of γ alumina or mullite. Imogolite yields by DTA an endothermic peak at 390 to 420°C (663 to 693 K) because of dehydroxylation (Yoshinaga and Aomine, 1962).

Allophane and imogolite occur mostly in soils of volcanic origin. These soils have been classified under different names in the past, e.g., Andosols, humic allophane soils, Trumao soils, Kuroboku soils (Tan, 1964). They occur extensively in the continents and islands around the Pacific ocean, and have been found to a lesser extent also in the West Indies, Africa, Italy, and Australia. In the United States soils containing allophane are classified mostly as the Andepts (Inceptisols).

4.4 THE IDENTIFICATION OF CLAY MINERALS

In the preceding sections some of the methods for the identification of clays have been mentioned briefly without going into details on the techniques and physicochemical reactions involved. For a better comprehension it is perhaps necessary to discuss briefly three of the major methods often used, e.g., differential thermal analysis, x-ray diffraction analysis, and infrared spectroscopy.

Differential Thermal Analysis

This method, commonly referred to as DTA, is a widely used technique and is very useful especially in the identification of amorphous material when x-ray diffraction analysis yields only featureless curves (Tan and Hajek, 1977). It found application first in geology, and has been later extended to research and analysis in ceramics, glass, polymer, cement, plaster industries, and so forth.

Differential thermal analysis measures the differences in temperature developed between an unknown and a reference sample, as the two are heated side by side at a controlled heating rate from 0 to 1000°C. The reference material, also called *standard material*, is a substance that is thermally inert over the temperature range under investigation. A number of compounds have been used as standard sample, e.g., calcined Al_2O_3 and calcinod kaolinite (heated at 1000°C). The heating must be controlled at a uniform and steady rate through the analysis. Heating rates may vary from 0.1 to 2000°C/min. For most purposes a heating rate of 20°C/min is used. During the heating process, the unknown sample undergoes a thermal reaction and transformation. The latter is reflected by a difference in temperature between the unknown and reference sample. This difference in temperature is plotted in a graph, usually against the temperature at which the difference occurs (see Figure 4.32). If the temperature of the unknown sample becomes lower than that of the reference material, ΔT is negative, an endothermic peak is produced. When the temperature of the sample becomes higher than that of the reference material, ΔT is positive, an exothermic peak develops. The portion of the curve for which ΔT = zero (no difference in temperature between unknown and reference sample) is considered the base line. Ideally the base line is a straight line. Upon analysis by DTA, the mineral may undergo several thermal reactions, culminating in one or a series of endo- and exothermic peaks. The curve with the peaks serves as a fingerprint, while the specific temperatures at which the peaks develop are diagnostic for the identification of the mineral. In addition, the peak height or peak area of the main endothermic reaction can be used for quantitative determination.

Generally DTA can be performed with liquid or solid samples. With soil samples, whole soil, sand, silt, or clay fractions can be used. When whole soils are analyzed, the <2 mm fraction should be treated first with 30% H_2O_2 to remove organic matter, which may interfere by giving strong exothermic reactions. In general analysis

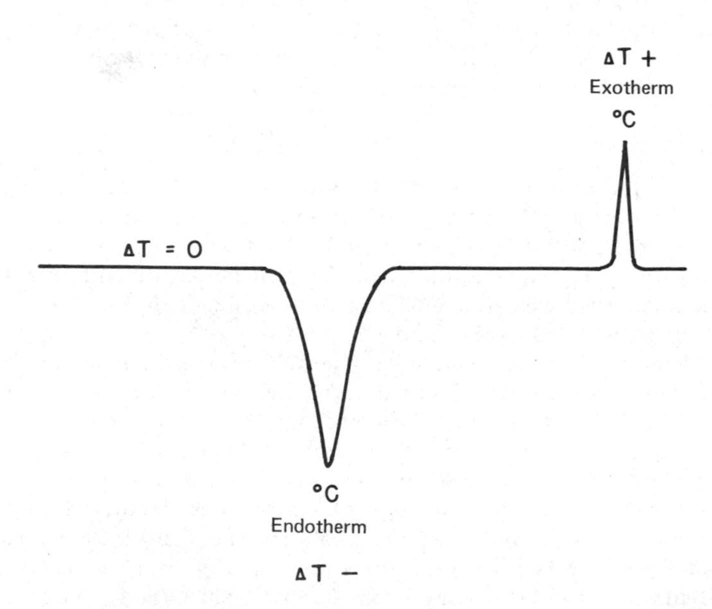

Figure 4.32 Idealized DTA curve.

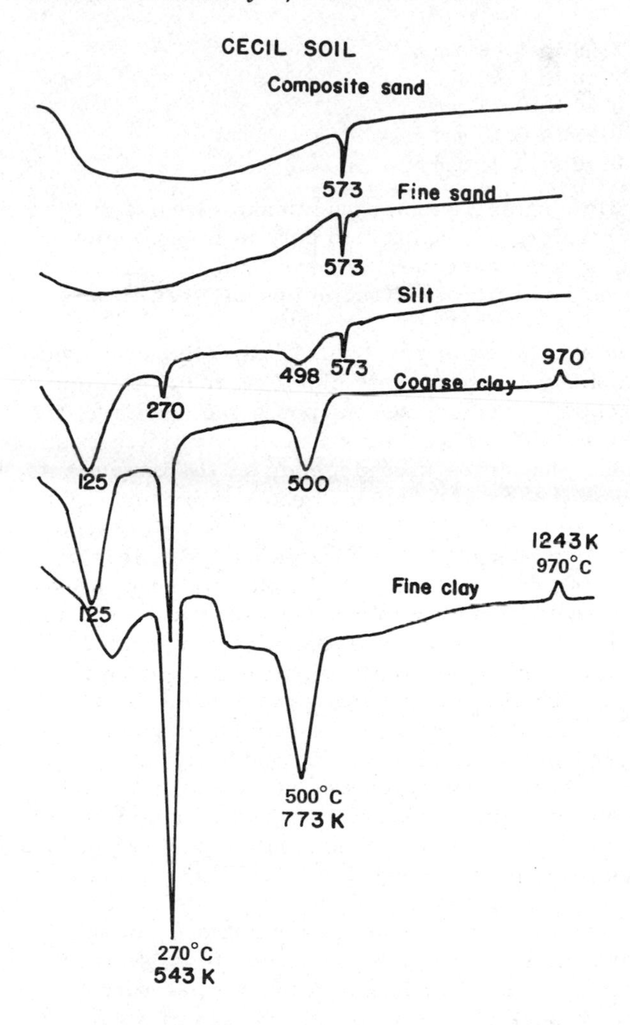

Figure 4.33 Differential thermal analysis (DTA) curves of composite sand (2.0 to 0.05 mm), fine sand (0.25 to 0.10 mm), silt (0.050 to 0.002 mm), coarse clay (0.002 to 0.0002 mm), and fine clay fractions (<0.2 μm) of a Cecil soil.

of whole soils gives only peaks of low intensity. These same peaks are very large and intense if the clay fractions are analyzed (see Figure 4.33). However, the quartz inversion peak at 573°C (864 K) is often absent in DTA curves of clays. The sand can be analyzed using the total sand fraction (2.0 to 0.05 mm) or one of the following sand fractions:

Very coarse sand	2.00 to 1.00 mm
Coarse sand	1.00 to 0.50 mm
Medium sand	0.50 to 0.25 mm
Fine sand	0.25 to 0.10 mm
Very fine sand	0.10 to 0.05 mm

DTA of sand shows mostly a strong endothermic peak of quartz at 573°C (846 K). It is, therefore, of importance only in investigation of primary minerals and/or iron-manganese concretions.

DTA of silt yields curves with peak resolutions between those of sand and whole soils.

The clay fraction of soils (<2 μm) can be used directly, or can be separated first, prior to analysis, into coarse clay (2.0 to 0.2 μm) and fine clay (<0.2 μm) fractions. For general purposes the clay fraction <2 μm gives satisfactory results for qualitative and quantitative interpretations. The amount of clay to be used depends on the instruments used. Instruments equipped with well holders need approximately 10 to 100 mg, whereas those equipped with Pt cups placed on ring-type thermocouples need only 1 to 10 mg. In qualitative analysis, it is often not necessary to weigh the sample for DTA, although comparison of curves should be made with curves obtained from identical amounts of samples. On the other hand, in quantitative analysis, the amount of sample must be weighed accurately, since height or area of main endothermic peak increases proportionally with sample size.

Qualitative identification of minerals can be achieved by using the DTA curves as fingerprints and comparing or matching them with DTA curves of standard minerals or with curves of well-known established minerals. Each mineral exhibits specific thermal reaction features (see Figure 4.34.). DTA curves of kaolinite are characterized by a strong endothermic peak at 450 to 600°C and by a strong exothermic peak at 900 to 1000°C. The endothermic peak is caused by dehydroxylation, whereas the exothermic peak is attributed to formation of γ alumina and/or mullite. The curve of halloysite is almost similar to that of kaolinite, but has in addition a low-temperature (100 to 200°C) endothermic peak of medium to strong intensity for loss of adsorbed interlayer water. Montmorillonite exhibits a DTA curve characterized by a low-temperature (100 to 200°C) endothermic peak, an endothermic peak between 600 and 750°C, and a small dip between 800 and 900°C followed by a weak exothermic peak between 900 and 1000°C.

Gibbsite and goethite are usually characterized by a strong endothermic peak only between 290 and 350°C. Often goethite and the other iron oxide minerals have their endothermic reaction at a higher temperature than gibbsite. Allophane exhibits DTA features with a strong low-temperature (50 to 150°C) endothermic peak and a strong exothermic peak at 900 to 1000°C. The low-temperature

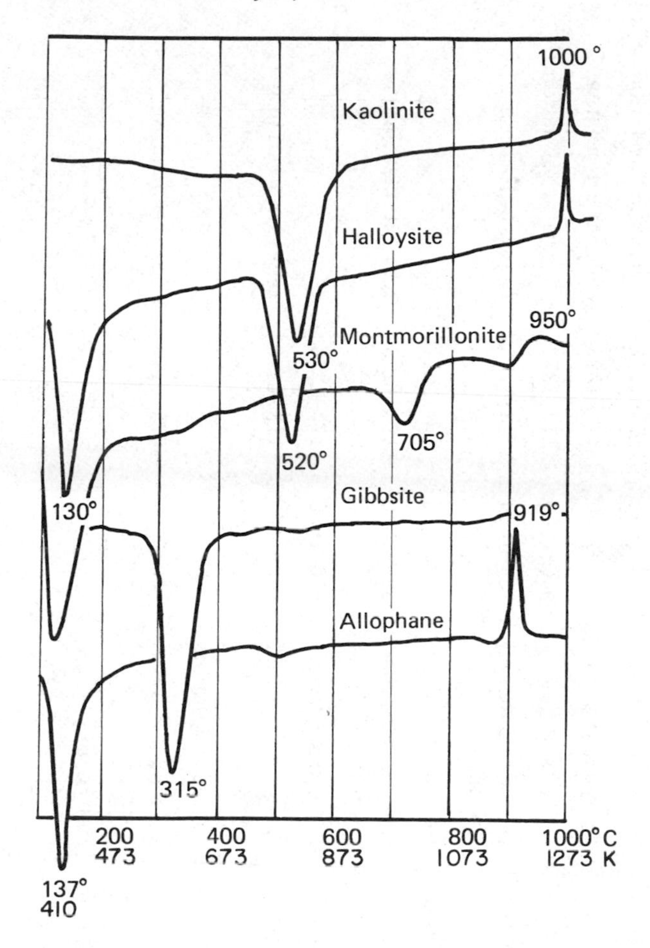

Figure 4.34 Characteristic differential thermal analysis (DTA) curves of selected clay minerals.

endothermal reaction is attributed to loss of adsorbed water, whereas the main exothermic reaction is caused by γ alumina formation.

A list of characteristic endo- and exothermic peaks of major clay minerals is provided in Table 4.9.

X-Ray Diffraction Analysis

This method is perhaps the most widely used method in the identification of clays. It is mainly for qualitative analysis, although frequently semiquantitative determination of clays has been carried out.

Table 4.9 DTA Endo- and Exothermic Peaks of Major Clay Minerals and the Reactions Causing the Peaks

Mineral	Endothermic peak		Main reaction	Exothermic peak		Main reaction
	°C	K		°C	K	
Kaolinite	500-600	773-873	Dehydroxylation	900-1000	1173-1273	γ alumina formation
Dickite	500-700	773-973	Dehydroxylation	900-1000	1173-1273	γ alumina formation
Nacrite	500-700	773-973	Dehydroxylation	900-1000	1173-1273	γ alumina formation
Montmorillonite	100-250	373-523	Lost of adsorbed water	900-1000	1173-1273	Recrystallization
	600-750	873-1023	Dehydroxylation			
Beidellite	100-250	373-523	Lost of adsorbed water	900-1000	1173-1273	Recrystallization
	500-600	773-873	Dehydroxylation	900-1000	1173-1273	
Nontronite	100-200	373-473	Lost of adsorbed water	9000-1000	1173-1273	Recrystallization
	500	773	Dehydroxylation			
Vermiculite	150	423	Lost of adsorbed water	800-900	1073-1173	Recrystallization
	850	1123	Dehydroxylation			

Illite	100-200	373-473	Lost of adsorbed water	920-950	1193-1223	Recrystallization
	600	873	Dehydroxylation			
	900-920	1173-1193	Dehydroxylation			
Chlorite	500-600	773-873	Dehydroxylation	800	1073	
Halloysite	100-200	373-473	Lost of adsorbed water	900-1000	1173-1273	γ alumina formation
	500-600	773-873	Dehydroxylation			
Gibbsite	250-350	523-623	Dehydroxylation			
Boehmite	570	843	Dehydroxylation			
Diaspore	400-500	673-773	Dehydroxylation			
Goethite	300-400	573-673	Dehydroxylation			
Quartz	—	—	—	573	843	α- to-β inversion
Allophane	50-150	323-432	Lost of adsorbed water	800-900	1073-1173	γ alumina formation

Source: Mackenzie (1975), Mackenzie and Caillere (1975), and Tan and Hajek (1977).

X-ray diffraction analysis is a nondestructive method, meaning that the sample is not affected by the analysis, and can be used for other analyses. However, the method is not applicable to analysis of amorphous or noncrystalline materials.

The basis for the use of x-rays in the investigation of soil clays is the systematic arrangement of atoms or ions in crystal planes as discussed previously. Each mineral species is characterized by a specific atomic arrangement, creating characteristic atomic planes, which can diffract (reflect) x-rays. X-rays are electromagnetic radiation of short wavelength. In most crystals, the atomic spacings, or crystal planes, have almost the same dimension as the wavelength of x-rays. Laue was perhaps the first to discover, in 1912, that x-rays can be diffracted by the atoms in a crystal plane, producing characteristic patterns when recorded. This diffraction pattern is used as a fingerprint in the identification of mineral species.

X-rays are produced in a x-ray tube by fast moving electrons hitting a metal target. The excited atoms in the target emit radiation with a wavelength between 0.01 and 100 Å, the wavelength of Kα and Kβ radiation. Most metals emit wide bands of Kα and Kβ radiation, e.g., Cu target. By using a nickel filter, the Cu Kβ radiation can be blocked or adsorbed and the Cu Kα radiation is then isolated for use in the analysis. If a beam of Cu Kα radiation hits a crystal plane of a mineral (see Figure 4.35), the x-rays are scattered by the atoms of the crystal. To have diffraction occurring, reinforcement must take place of the scattered x-rays in a definite direction. Reinforcement of scattered x-rays becomes quantitative only if Bragg's law is obeyed. Bragg's law is defined as follows:

$$n \lambda = 2d \sin \theta$$

where

- d = spacings between atomic planes in the crystal
- λ = wavelength
- θ = glancing angle of diffraction
- n = order of diffraction

The true lattice spacing for the (001) basal plane is the d (001) or the d (hkl) spacing.

Bragg's law predicts that all planes in a crystal diffract x-rays, when the crystal is inclined at certain angles to the incident beam. The angles θ depend on the wavelength λ and on d. If Figure 4.35 is studied, the incident beam DEF has traveled several integral numbers of wavelength ($n\lambda$) farther than incident beam ABC. Diffraction from a succession of equally spaced atomic planes yields a diffraction maximum. If these diffractions are received by a photographic film, a series of spots or lines (bands) are produced. The position of the lines is related directly to the d spacings.

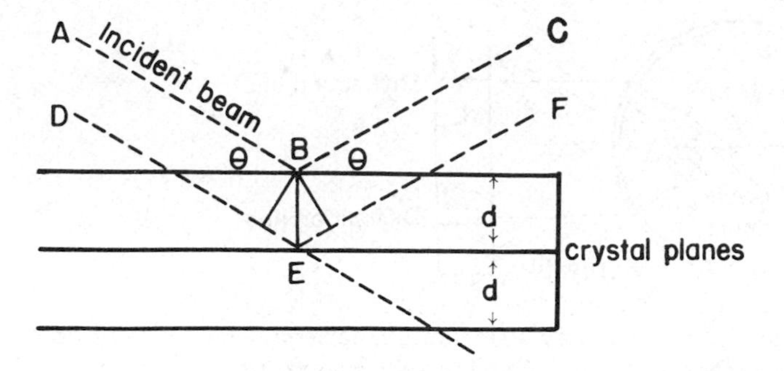

Figure 4.35 Schematic drawing of an x-ray incident beam diffracting from crystal planes, obeying Bragg's law: $n\lambda = 2d\ \theta$.

The value d(001)/n, called *lattice spacing*, can be measured from results of x-ray diffraction analysis. If n = 1, the d(001)/1 represents the first-order diffraction spacing. If n = 2, then d(001)/2 represents the second-order diffraction spacing. The series of d/n values obtained, together with the intensity of the x-ray diffraction peaks, are diagnostic for identification of mineral species.

The accepted unit for lattice spacing is the angström unit (10^{-10} m = 1Å), which corresponds to the unit of x-ray wavelength. In some books, the nanometer (1 Å = 0.1 nm) is preferred, while in older books, the kX unit is used. The kX unit is based on the effective spacing of cleavage planes of calcite (= 3.0290 kX). Conversion of kX units into Å units is easily conducted by multiplying with 1.0020.

The samples for x-ray diffraction analysis can be prepared as a random powder sample or as an oriented sample. In a random powder sample, the crystals lie in a random position to each others. With the aid of glycerol and gum tragacanth the clay sample is made into a paste, and rolled into a rod of 0.3 to 0.5 mm thickness. Powder samples can also be prepared by pushing the paste into specially designed wedge holders. The random powder samples are usually analysed by a powder camera x-ray unit (see Figure 4.36). As indicated previously, the diffracted x-rays produce lines or bands on a photographic film. The position of the lines corresponds to the d spacings of the crystal planes of the mineral.

A currently more popular method of mounting samples for x-ray analysis is the preparation of oriented samples on microscopic glass slide, or on porous ceramic plates. A clay suspension is made properly and pipetted onto the slide, so that approximately 15 to 25 mg of clay is transferred per 10 cm^2. After the sample has been allowed to dry

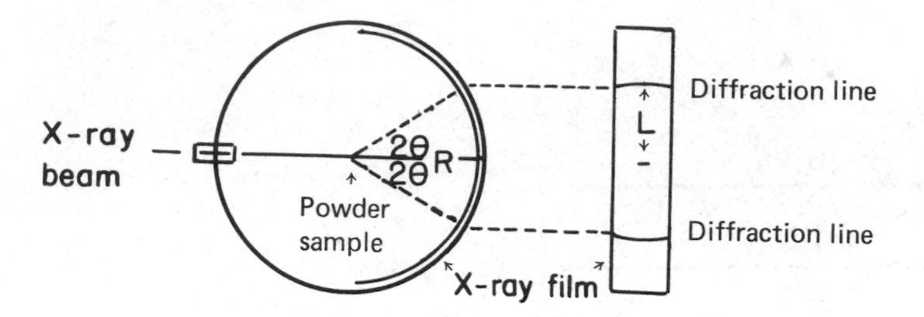

Figure 4.36 Random powder x-ray diffraction camera. R = radius of camera, and L = distance between location of primary beam and diffraction maximum. L is calculated as follows: $L/2\pi R = 2\theta/360°$.

at room temperature, it is ready for analysis with a direct recording x-ray spectrometer, in which the x-ray patterns are printed on charts. The results are normally shown in terms of 2θ values. However, a number of tables are available to convert the 2θ values into d spacings units. Generally clay minerals exhibit d spacings in the range of 30 to 3 Å, which corresponds to 2 to 30° 2θ angles. Highest intensity of diffraction maxima is obtained from d (001) planes. The first-order d (001) diffraction peak, and in many cases together with the second order diffraction peak, are diagnostic for the identification of the mineral species. The following illustrations serve as a few examples.

Kaolinite (see Figures 4.37 to 4.39) exhibits characteristic first-order diffraction at an angle of $2\theta = 12.4°$. The latter corresponds after conversion to a d (001) spacing of 7.13 Å. The second-order diffraction is at $2\theta = 25.0°$, which corresponds to a d spacing of 3.56 Å.

Montmorillonite (air dry) is characterized by a first-order x-ray diffraction peak of 12.3 Å, which shifts to 17.7 Å after solvation of the sample. The second-order diffraction peak is usually absent.

Illite exhibits a first-order diffraction peak of 10.1 Å. This peak will not collapse or shift after potassium and magnesium saturation or solvation of the sample.

Gibbsite is identified by the dominant diffraction at approxitely $2\theta = 18.4°$, corresponding to a d spacing of 4.82 Å.

Goethite is easily recognized from the dominant peak at approximately $2\theta = 21.6°$, corresponding to a d spacing of 4.12 Å.

For additional characteristic d spacing values of other clay minerals, the following list can be used as a reference (see Table 4.10).

As can be noticed from the d spacings as listed in Table 4.10, a number of the minerals have similar, or overlapping, diffraction peaks. In such cases pretreatments of samples are required to

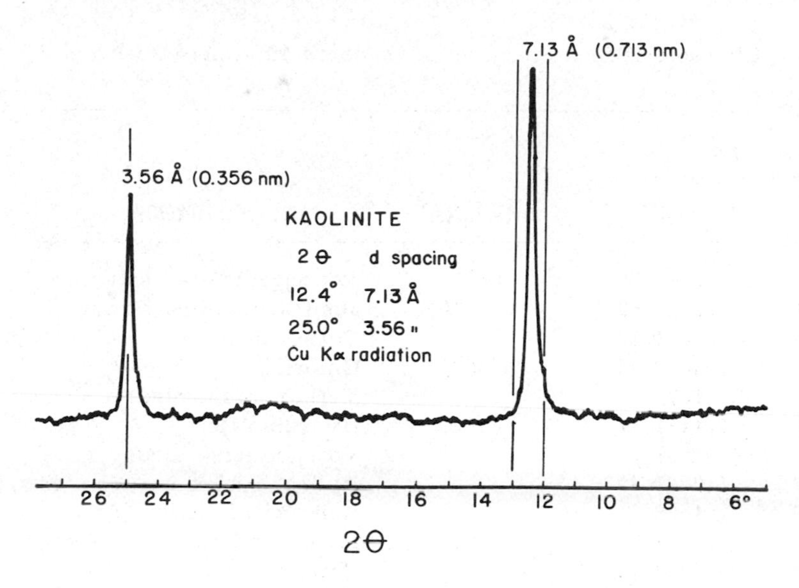

Figure 4.37 Interpretation and identification of kaolinite using its x-ray diffraction pattern. A 12.4 2θ reading corresponds to a d spacing of 7.13 Å (0.713 nm), which is the diagnostic d spacing of kaolinite. See Appendix D for 2θ-d spacings values.

distinguish these minerals. Four major methods frequently used for pretreating the samples prior to analysis are (1) K saturation, (2) Mg saturation, (3) solvation of Mg-saturated samples, and (4) heating at 500°C. The first three methods stated have the purpose of distinguishing between expanding and nonexpanding minerals (see Table 4.11). Potassium saturation will normally effect a collapse of intermicellar spacings, and d spacings of 20 to 17 Å as exhibited by expanding montmorillonites may collapse to 10 Å. Reconstitution of the d spacing to 17 Å can be achieved by solvation. None of these treatments will have any effect on the d spacings of nonexpanding minerals, e.g., kaolinite. Heating at 500°C is usually done to distinguish between vermiculite, chlorite and kaolinite. Vermiculite may have interlayer hydroxy alumina complexes in its structure. These interlayers will be destroyed by heating at 500°C, with the consequent collapse of the d spacing of vermiculite from 14 Å to 10 Å. On the other hand, heating at 500°C will show no effect on chlorite but will, however, produce a 7.2-Å (second-order) diffraction peak. The latter can be confused for kaolinite. However, heating at 500°C will make kaolinite amorphous to x-ray diffraction, and both the 7.13 and 3.56 Å peaks of kaolinite will disappear.

Table 4.10 Characteristic d Spacings of Selected Minerals (Cu Kα Radiation)

d spacings			
Å	nm	Intensity[a]	Clay mineral
17.7-17.0	1.77-1.70	(10)	Montmorillonite, solvated
12.0-15.0	1.20-1.50	(8)-(10)	Montmorillonite, airdry
14.0-15.0	1.40-1.50	(1)	Vermiculite
13.0-14.0	1.30-1.40	(3)-(8)	Chlorites
12.2	1.22		Vermiculites, airdry
11.4-11.7	1.14-1.17	(10)	Hydrobiotite
10.7	1.07		Vermiculites, airdry
10.0-14.0	1.00-1.40		Hydrous mica
10.8	1.08	(10)	Halloysite, hydrated
9.0-10.0	0.90-1.00	(10)	Illites and micas
7.2-7.5	0.72-0.75	(8)	Metahalloysite
7.1-7.2	0.71-0.72	(10)	Kaolinite, dickite, nacrite
6.44	0.644	(6)	Palygorskite
5.90	0.59	(0)-(3)	Montmorillonite, solvated
5.42	0.542	(5)	Palygorskite
5.00	0.500	(9)	Muscovite
4.7-4.8	0.47-0.48	(9)	Chlorite
4.6	0.46	(5)	Vermiculite
4.6	0.46	(10)	Sepiolite
4.4-4.5	0.44-0.45	(9)	Illite, muscovite
4.49	0.449	(8)	Palygorskite
4.45-4.46	0.445-0.446	(4)	Kaolinite
		(6)	Dickite
		(8)	Fireclay
4.43	0.443	(6)	Dickite
4.42	0.442	(10)	Metahalloysite
4.40	0.440	(8)	Nacrite
4.35-4.36	0.435-0.436	(6)	Kaolinite, dickite
4.26	0.426	(4)	Dickite
4.20-4.30	0.420-0.430	(5)	Palygorskite
4.17	0.417	(6)	Kaolinite
4.13	0.413	(6)	Dickite
4.12	0.412	(3)	Kaolinite
3.84	0.384	(4)	Kaolinite
3.82	0.382	(5)	Sepiolite
3.78	0.378	(6)	Dickite
3.69	0.369	(5)	Palygorskite

Table 4.10 (Continued)

d spacings			
Å	nm	Intensity[a]	Clay mineral
3.56-3.58	0.356-0.358	(10)	Kaolinite
		(10)	Dickite
		(9)	Nacrite
		(8)	Metahalloysite
		Silicates	
9.20-9.40	0.920-0.940	(9)	Talc
9.10-9.20	0.910-0.920	(6)	Pyrophyllite
7.10-7.20	0.710-0.720	(6)	Antigorite
7.10-7.20	0.710-0.720	(6)	Chrysotile
6.30-6.45	0.630 0.645	(4)-(6)	Feldspars
5.40	0.540	(7)	Mullite
4.60-4.70	0.460-0.470	(6)	Talc
4.57	0.457	(5)	Prophyllite
4.00-4.20	0.400-0.420	(8)	Feldspars
3.80-3.90	0.380-0.390	(2)-(7)	Feldspars
3.73-3.75	0.373-0.375	(4)-(8)	Feldspars
3.64-3.67	0.364-0.367	(3)-(8)	Feldspars
3.59-3.60	0.359-0.360	(7)	Antigorite
		(6)	Chrysotile
3.44-3.48	0.344-0.348	(3)-(6)	Feldspars
3.39	0.339	(10)	Mullite
3.36	0.336	(3)	Pyrophyllite
3.10-3.25	0.310-0.325	(7)-(10)	Feldspars
		Oxides and Hydroxydes	
6.25	0.625	(10)	Lepidocrocite
6.23	0.623	(10)	Boehmite
4.96	0.496	(3)	Goethite
4.85	0.485	(3)	Magnetite
4.83	0.483	(10)	Gibbsite
4.72	0.472	(10)	Bayerite
4.62	0.462	(8)	Spinel
4.36	0.436	(8)	Bayerite
4.34	0.434	(6)	Gibbsite
4.29	0.429	(10)	Gypsum
4.21	0.421	(7)	Quartz

Table 4.10 (Continued)

d spacings			
Å	nm	Intensity[a]	Clay mineral
4.15	0.415	(10)	Goethite
4.05	0.405	(10)	Crystoballite
3.98	0.398	(10)	Diaspore
3.84	0.384	(6)	Calcite
3.73	0.373	(7)	Ilmenite
3.72	0.372	(3)	Maghemite
3.67	0.367	(7)	Heamatite
3.36	0.336	(3)	Goethite
3.35	0.335	(10)	Quartz
3.30	0.330	(3)	Gibbsite
3.28	0.328	(9)	Lepidocrocite
3.20	0.320	(6)	Bayerite
3.16	0.316	(10)	Boehmite
3.15	0.315	(4)	Crystoballite
3.06	0.306	(7)	Gypsum
3.30	0.330	(10)	Calcite

[a]Numbers in parentheses refer to intensity on a scale of 1 to 10 (with 1 equaling weak).

Infrared Spectroscopy

Recently infrared spectroscopy has found extensive applications in clay mineralogy studies. Amorphous as well as crystalline clays absorb infrared radiation, and the method can be used when x-ray analysis makes identification difficult. The infrared absorption spectrum of a mineral has a characteristic pattern, which not only permits the identification of the mineral, but also reveals the presence of major functional groups within the structure of the particular compound under investigation.

Infrared absorption is related to molecular or atomic vibrations, and only radiation with a similar frequency as that of the vibration will be absorbed. Atoms and molecules within a compound oscillate or vibrate with frequencies of approximately 10^{13} to 10^{14} cps. These frequencies correspond to the frequencies of infrared radiation, and infrared radiation can, therefore, be absorbed by molecular vibrations, when the interaction is accompanied by a change in dipole moment. A rapid vibration of atoms yields a rapid change in dipole moment, and absorption of infrared radiation is intense. On the other

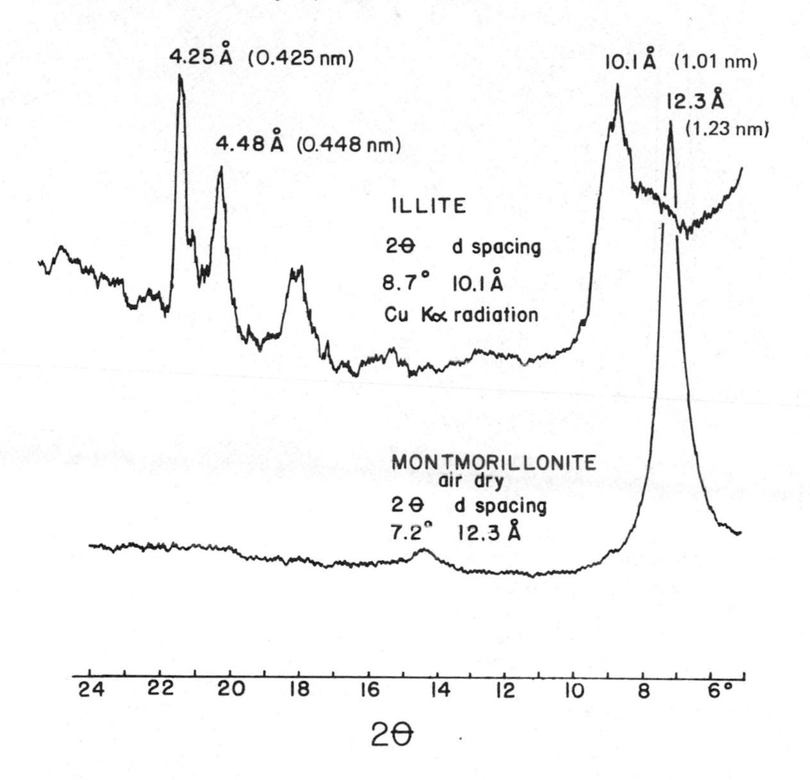

Figure 4.38 Interpretation and identification of illite and montmorillonite using their x-ray diffraction pattern. (Top) A 8.7 2θ reading corresponds to a d spacing of 10.1 Å (1.01 nm), which is diagnostic for illite. (Bottom) A 7.2 2θ reading corresponds to a d spacing of 12.3 Å, which is characteristic for airdry montmorillonite. See Appendix D for 2θ-d spacings values.

hand, a weak vibration of atoms produces a slow change in dipole, and consequently absorption of infrared radiation is relatively weak. Symmetrical molecules will also often not absorb infrared radiation.

Molecular or atomic vibrations cause the interatomic distance to change because the atomic movement, called *oscillation*, subjects the atoms to a periodic displacement with respect to one another. The frequency of vibration obeys the law of simple harmonic motion formulated as

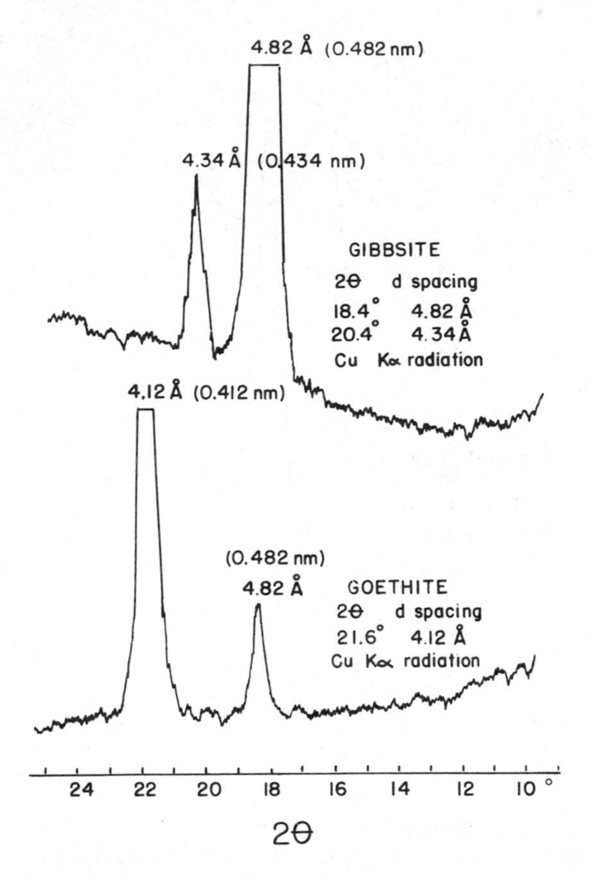

Figure 4.39 Interpretation and identification of gibbsite and goethite using their x-ray diffraction patterns. (Top) A 18.4 2θ reading corresponds to a d spacing of 4.82 Å (0.482 nm), which is the major diagnostic peak for gibbsite. (Bottom) A 21.6 2θ reading corresponds to a d spacing of 4.12 Å (0.412 nm), characteristic for goethite.

$$V = \frac{1}{2\pi c}\sqrt{\frac{k}{m}}$$

where

V = frequency of vibration in cm^{-1}
c = velocity of light in vacuum
m = reduced mass of the vibrating atoms
k = force constant in dyn/cm

Two types of vibrations are distinguished, e.g., (1) stretching vibrations, or deformation, in which the atoms are oscillating in the

Table 4.11 Effect of Pretreatments on d Spacings of Selected Clay Minerals

d spacings Å	nm	Minerals (air-dry)
		K-saturated samples
14	1.4	Vermiculites, chlorites
10-12	1.0-1.2	Montmorillonites, illites
7.2-7.5	0.72-0.75	Halloysites, metahalloysites
7.15	0.715	Kaolinites, chlorites
		Mg-saturated samples
14	1.40	Vermiculites, chlorites, montmorillonites and illites
10-12	1.00-1.20	Illites, halloysites
7.2-7.5	0.72-0.75	Kaolinites, chlorites
		Solvated Mg-saturated samples
17-18	1.70-1.80	Montmorillonites
14	1.40	Vermiculites, chlorites
10-12	1.00-1.20	Illites, halloysites
7.15	0.715	Kaolinites
		Heated at 500°C (773 K)
14	1.40	Chlorites
10	1.00	Vermiculites
7.0	0.70	Chlorites (kaolinite becomes amorphous)

direction of the bond axis without changing bond angles, and (2) bending vibrations, in which the movement of atoms produces a change in bond angles. The restoring force acting on stretching vibrations is usually greater than that required to restore bending vibrations. Therefore, stretching vibrations occur at higher frequencies than bending vibrations. The highest frequencies observed in minerals are those of the stretching vibrations of hydroxyl, OH, groups which occur between 3700 and 2000 cm^{-1}. Bending vibrations occur at lower frequencies, from 1630 to 400 cm^{-1}.

Liquid, gas, and solid samples can be used in infrared spectroscopy. Liquid samples are usually pipetted or injected into infrared cells provided with a NaCl or KBr crystal window. Gas samples are also introduced in cells, similar to the cells stated above. Infared gas cells are larger than cells for liquid samples and ensure better interaction between infrared radiation and the gas by providing longer path length. Both NaCl and KBr are infrared inactive and will not interfere in the analysis.

Solid samples should be ground to approximately <2 μm, since coarse particles tend to produce scattering of infrared radiation. Clay fractions (<2 μm) separated by mechanical analysis can be used directly, or can be separated first by centrifugation into coarse (2.0 to 0.2 μm) and fine (<0.2 μm) clay fractions. If grinding is necessary, it should be carried out with care, since vigorous grinding tends to destroy the mineral structure (becomes amorphous) and tends to increase the hygroscopic nature of the sample.

Several methods have been proposed for mounting solid samples in infrared analysis, e.g., (1) mull method, (2) KBr pellet technique or (3) clay film technique on demountable cells and/or other support material. The most widely used method is the KBr pellet technique, by which a weighed sample (1 to 10 mg) is carefully ground with 100 mg KBr, and pressed into a clear transparent pellet. Sodium chloride also appears to be suitable for use in the pellet method. Currently, the use of clay films has attracted considerable attention. The present author noticed that clay films prepared on infrared cells, such as Irtran-II (ZnS crystal), or NaCl crystals, give better infrared resolutions than clay samples mounted by the KBr pellet technique (see Figures 4.40 and 4.41). However, one disadvantage of the clay film technique is that it takes more time to prepare a clay film than to make a pellet.

In the clay film mounting method, clay or soil samples are made into a suspension by sonification. They are then pipetted onto Irtan-II window cells, so that 1 mg/cm^2 or 5 mg $clay/cm^2$ are transferred onto the cells. After drying at room temperature, the cells are scanned from 4000 to 600 cm^{-1} or lower.

The results of infrared analysis, as reflected by absorption curves, are using frequencies as the main units. The *frequency* V, also known as the *wavenumber*, is defined as the number of waves,

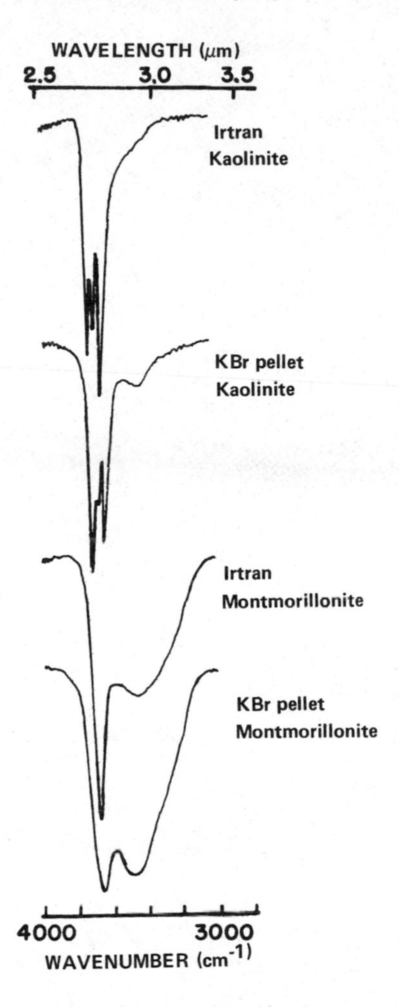

Figure 4.40 Characteristic infrared features in the group frequency region of kaolinite and montmorillonite.

or wavelengths, per centimeter (cm^{-1}). It is related to the wavelength λ as follows:

$$V = \frac{10^4}{\lambda}$$

in which V is expressed in cm^{-1}, and λ in micrometers (μm). The region often analyzed by infrared spectroscopy is in the range of 4000 to 600 cm^{-1} (equivalent to 2.5 to 25 μm) or lower. In many instances, the results of infrared analysis are recorded in the trans-

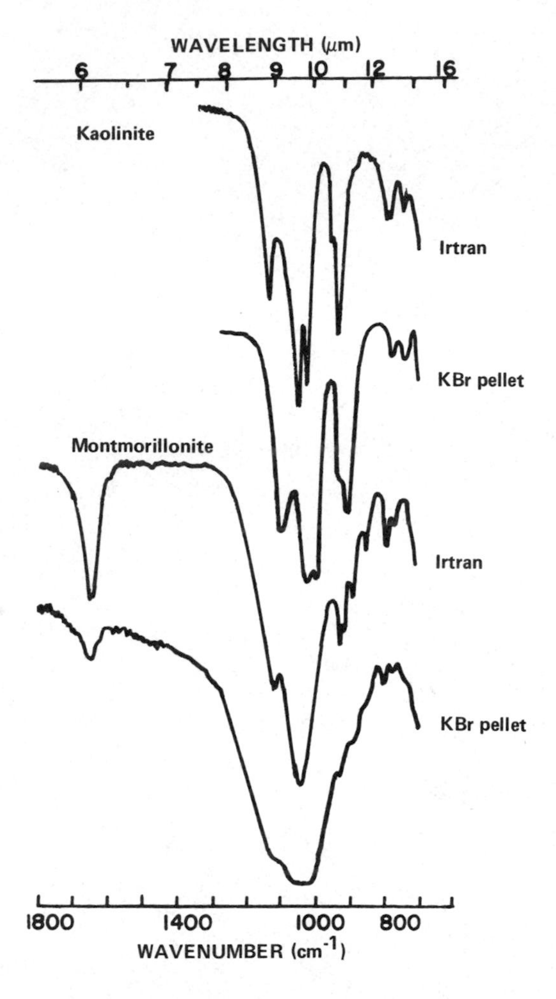

Figure 4.41 Characteristic infrared features in the fingerprint region of kaolinite and montmorillonite.

mittance mode. The latter then yields curves which have an upside down appearance when compared to absorption curves (Tan et al., 1978).

Two groups of frequency regions usually characterize the infrared curves of most clay minerals:

1. Region between 4000 and 3000 cm^{-1} attributed to stretching vibrations of adsorbed water and/or octahedral OH groups, called the *functional group region*.
2. Region between 1400 and 800 cm^{-1}, attributed to Al-OH and/or Si-O vibrations, called the *fingerprint region*.

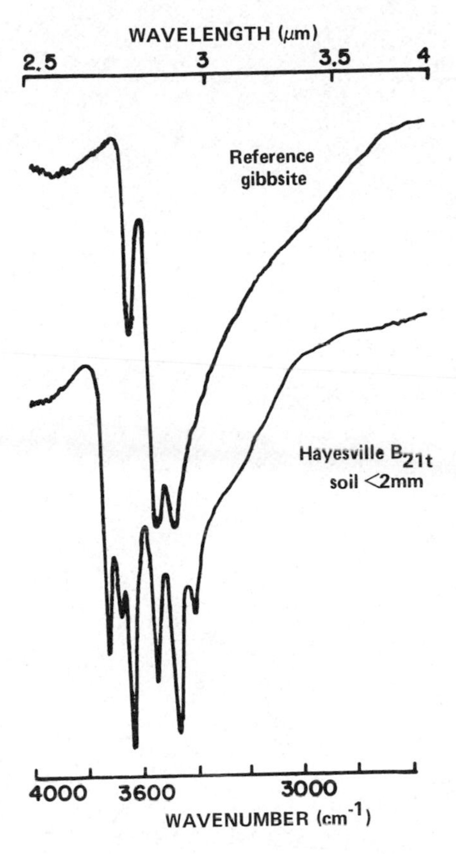

Figure 4.42 Characteristic infrared features in the group frequency region of reference gibbsite and a Hayesville (Ultisols) B_{21t} soil sample containing gibbsite.

The infrared curves of kaolinite, montmorillonite, and gibbsite, shown in Figures 4.40 to 4.43, serve as illustrations.

The infrared curve of kaolinite is usually characterized by two strong bands for octahedral OH stretching vibrations between 3800 and 3600 cm^{-1}, when the sample is mounted by the KBr pellet technique. An additional third and very sharp band is present at 3670 cm^{-1} when the samples are mounted as films on Irtran-II cells. The fingerprint region, or the lower frequency region, exhibits sharp bands for kaolinite at 1150, and 1080 cm^{-1} for O-Al-OH vibrations using clay films on Irtran windows. In addition a sharp 1020 cm^{-1} band for Si-O, and sharp bands at 910 to 920 cm^{-1} for Al-OH vibrations are present. Using KBr pellets, the bands at 1080 and 1020

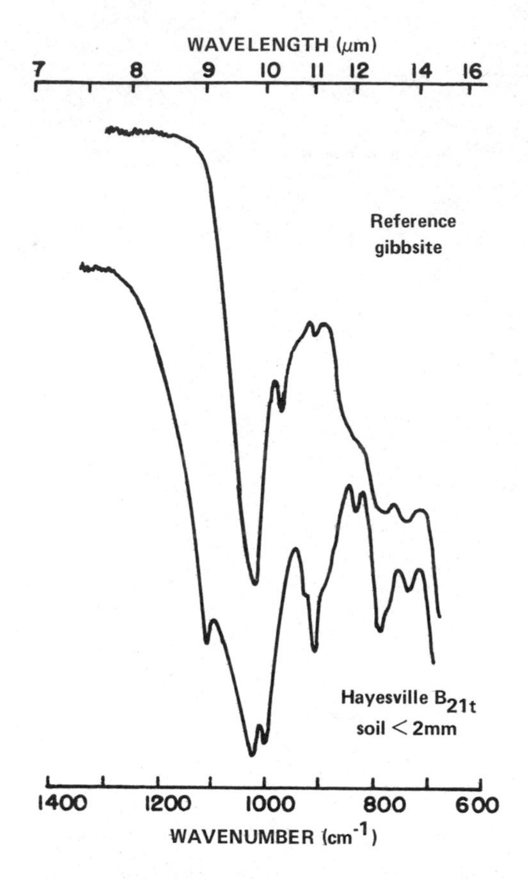

Figure 4.43 Characteristic infrared features in the fingerprint region of reference gibbsite and a Hayesville (Ultisols) B_{21t} soil sample containing gibbsite.

cm^{-1} appear only as a weakly segregated duplet in most analysis reported for kaolinite.

Montmorillonite also exhibits a better resolved curve with the clay film technique on Irtran-II. The KBr curve is characterized by one broad band followed by a water band, and one additional dominant broad band at 3640, 3420, and 1050 cm^{-1}, respectively. However, using clay films on Irtran-II, the band at 3640 cm^{-1} for OH-stretching vibrations and at 1050 cm^{-1} for Si-O vibrations become very strong and sharp. In addition bands at 1150, 910, 880, and 850 cm^{-1} also increase sharply in intensity in clay film samples.

Standard reference gibbsite (purchased from the Wards Scient. Establ. Co.) is characterized by an absorption band at 3620 cm^{-1}, and by a duplet at 3540 and at 3480 cm^{-1}. In the low frequency region, gibbsite shows only one dominant peak at 1030 cm^{-1} for O—Al—OH vibration. This peak is the reason for determining the kaolinite band at 1080 cm^{-1} as a separate independent band, rather than calling it a duplet together with the 1020 cm^{-1} band.

Soil gibbsite (see Hayesville B21t soil) exhibits an infrared curve characterized by a triplet between 3600 and 3400 cm^{-1}, instead of the duplet observed for the reference gibbsite. The 3620 cm^{-1} band is overlapping with that of the octahedral OH band of kaolinite in the Hayesville clay fraction. Apparently, the different infrared pattern suggests the presence of different types of gibbsite in soils.

4.5 SURFACE CHEMISTRY OF SOIL CLAYS

Many, if not all, of the chemical reactions of soil clays, are surface phenomena, e.g., cation exchange, adsorption of water. From the preceeding section on clay mineral structure, it follows that clay surfaces can be divided into at least three categories:

1. Surfaces formed mainly by Si-O-Si linkages of silica tetrahedrons
2. Surfaces formed by O-Al-OH linkages of alumina octahedrons
3. Surfaces formed by -Si-OH or -Al-OH of amorphous compounds

The first category of surfaces is characterized by surface planes of oxygen atoms, underlaid by silicon atoms of the tetrahedrons. The Si-O-Si bond is called a *siloxane bond* by Sticher and Bach (1966), and because of this, this type of surface can conveniently be called the siloxane surface, and is typical of 2:1 types of clay. The charge of a siloxane surface is mainly attributed to isomorphous substitution of the underlaying silicon atoms of the tetrahedrons.

The second type of clay surfaces is characterized by planes of exposed hydroxyl, OH, groups, underlaid by Al, Fe or Mg atoms in the center of the octahedrons. Because of the latter, it can perhaps be called the oxy-hydroxide surface. Kaolinite and other 1:1 types of clay usually have siloxane surfaces on one basal plane, and oxyhydroxide surfaces on other basal planes. The exposed hydroxyl groups are subject to dissociation and, therefore, play an important role in the development of negative charges.

The third type of surfaces is formed by -Si-OH, called *silanol surfaces*, and -Al-OH, called *aluminol surfaces*. They are typically present in soils containing large amounts of amorphous silica gel and/or allophane. The behavior of these surfaces is expected to be quite different from the other two types mentioned above. Usually, the compounds with silanol and aluminol surfaces have a very large surface area, while all the hydroxyl groups are easily accessible.

Table 4.12 Specific Surface Areas of Selected Clay Minerals

	Total surface area, m^2/g		
	H_2O method	CPB method	N_2 gas method
Montmorillonite	300	800	784
Mica-smectite (interstratified)	57	152	109
Kaolinite	17	15	32
Allophane	484	0	157

Source: Adapted from Greenland and Quirk (1962, 1964), Dixon (1977), and Wada (1977)

4.6 SURFACE AREAS

In connection with the surfaces of clay minerals is the problem of surface areas, needed for quantitative interpretation of surface properties in relation to soils or clay behavior. The surface area or specific surface area can be measured by two types of methods. The first method is electron microscopy, which is considered relatively the simplest method. The second type is a group of methods based on adsorption of compounds in the vapor or gas phase. Depending on the method used, a wide range of values is obtained for surface area. The data in Table 4.12 show some of the variations in surface areas according to adsorption of water, cetyl pyridinium bromide (CPB), or N_2 gas.

4.7 ORIGIN OF NEGATIVE CHARGE IN SOIL CLAYS

As indicated earlier, soil clays ordinarily carry an electronegative charge, which gives rise to cation exchange reactions. This charge is the result of one or more of several different reactions. Two major sources for the origin of negative charges are described below.

Isomorphous Substitution

This is believed to be a major source of negative charges in 2:1 layer clays. Part of the silicon in the tetrahedral layer is subject to replacement by ions of similar size, usually Al^{3+}. In the same manner, part of the Al in the octahedral sheet may be replaced by Mg^{2+}, without disturbing the crystal structure. Such a process of replacement is called *isomorphous substitution*. The resulting negative charge is considered a permanent charge since it will not change with changing pH.

The ease with which isomorphous substitution takes place depends on the size and valence of the ions involved. It occurs only with ions of comparable size. The difference in dimensions of substituted ions was reported to be no more than 15%, and the valency between those substituted ions should not differ more than one unit (Paton, 1978). In Figure 4.44, the relative dimensions of a few selected ions are given as illustrations. It can be observed that Na^+ and Ca^{2+} are almost of equal size and can replace one another with relative ease, in spite of the larger valence of Ca^{2+}. Potassium is expected to be unable to replace Na^+ or Ca^{2+} since it is approximately 1.4 times larger than the latter two ions. Magnesium and the iron ions are also of almost equal sizes, and may substitute for each other, the sizes being within 15% difference of one another. Aluminum is in size between Si^{4+} and Mg^{2+} or Fe^{3+} and is capable of replacing with varying degrees of ease any of these (Paton, 1978).

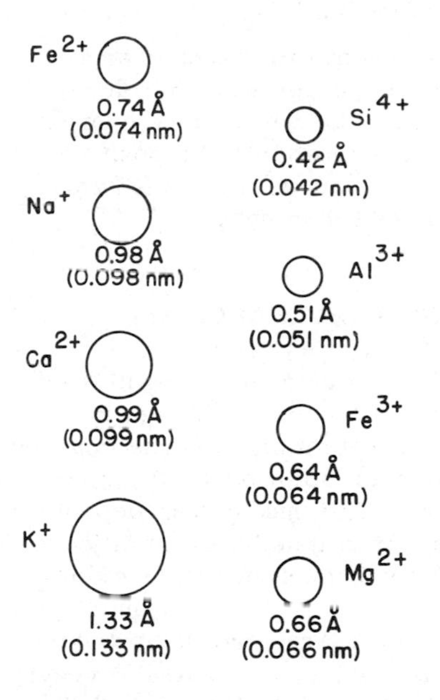

Figure 4.44 Relative dimensions of selected ions commonly found in soils.

Dissociation of Exposed Hydroxyl Groups

The appearance of OH groups on crystal edges or on exposed planes as discussed earlier can also give rise to negative charges. Especially at high pH, the hydrogen of these hydroxyls dissociates slightly and the surface of the clay is left with the negative charge of the oxygen ions. This type of negative charge is called *variable* or *pH-dependent charge.* The magnitude of the variable charge varies with pH and type of colloid. It is an important type of charge for 1:1 layer, iron and aluminum oxide clays, and organic colloids. As pointed out earlier, kaolinite also has subbasal hydroxyl groups. Since the latter are surrounded by a network of oxygen atoms from the silica tetrahedrons, it is expected that the dissociation and consequent contribution to negative charges from subbasal hydroxyl groups may be relatively small.

Not only can protons be dissociated from exposed OH groups, but the latter can also adsorb or gain protons. This process, important only in strongly acid media, creates positive charges. The reactions for dissociation and association of protons can be illustrated as follows:

Alkaline medium: $-Al\text{-}OH + OH^- \rightleftharpoons -Al\text{-}O^- + H_2O$

Acid medium: $-Al\text{-}OH + H^+ \rightleftharpoons -Al\text{-}OH_2^+$

The H^+ and OH^- ions, causing the development of surface charges, are also responsible for the electric surface potential. Therefore, they are called potential determining ions. The net surface charge will become zero if the negative charge density equals the positive charge density. The pH, at which the latter occurs, is called the *isoelectric point* or the *zero point charge* of the mineral.

4.8 POSITIVE CHARGES AND ZERO POINT OF CHARGE

Soil colloids may also exhibit positive charges as well as negative charges. The positive charges make possible anion exchange reactions and are very important in phosphate retention. These charges are thought to arise from the protonation or addition of H^+ ions to hydroxyl groups, as indicated above. This mechanism depends on pH and the valence of the metal ions. It is usually of significance in Al and Fe oxide clays, but it is of lesser importance in Si-oxides. For example, gibbsite is positively charged at pH 7.0 or lower.

Gibbsite and other soil colloids may be characterized by a particular pH at which the surface charge is electrically neutral. Previously, this point, or pH value, is called *zero point of charge* pH_o. At pH values above pH_o, the colloid is negatively charged. At pH values below pH_o, the colloid is positively charged.

The zero point of charge (ZPC) can be determined by either the titration method or by analysis of the amounts of adsorbed cations and anions as a function of pH and concentrations. If H^+ and OH^- ions are the main potential determining ions, ZPC is usually found by potentiometric titration and calculations using the formula:

$$\sigma_o = F\,(\Gamma_{H+} - \Gamma_{OH-})$$

where

σ_o = surface charge density

F = Faraday constant

Γ_{H^+}, Γ_{OH^-} = adsorption densities of H^+ and OH^-, respectively, in mEq/gram

The titrations are conducted with an indifferent electrolyte using several concentrations. Figure 4.45 is an example of presentation of results.

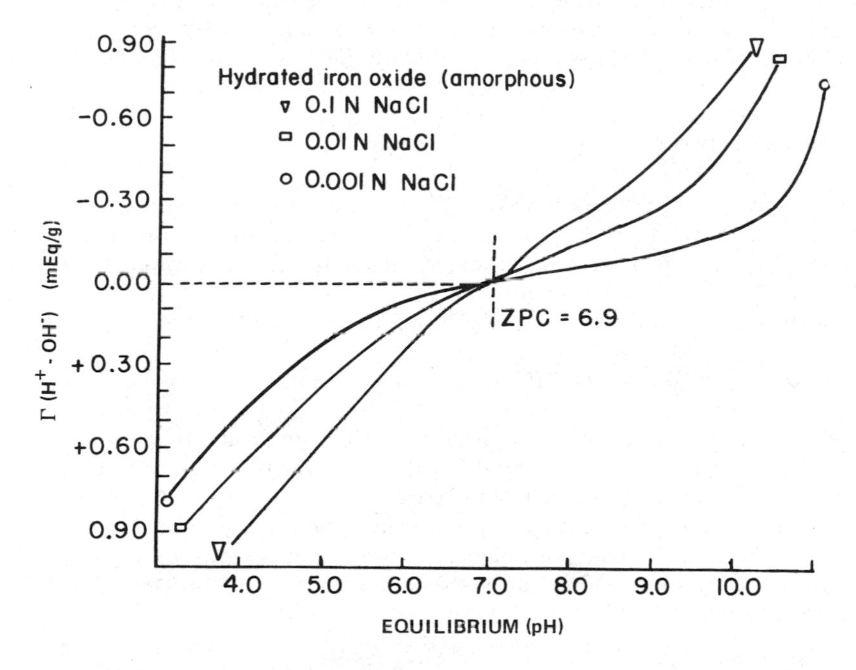

Figure 4.45 Potentiometric titration using NaCl at different strength in the determination of ZPC values of amorphous hydrated iron oxide. [From Gast (1977). Reproduced by permission of the Soil Science Society of America.]

4.9 THE USE OF ΔpH IN THE DETERMINATION OF NEGATIVE OR POSITIVE CHARGES

A relatively simple method to determine whether the net charge of the soil colloids is negative, zero, or positive, is the analysis of soil pH in 1 N KCl and in water. The difference between the two pH values is called ΔpH, and has been used in soil surveys for soil characterization:

$$\Delta pH = pH_{H_2O} - pH_{KCl}$$

The value for ΔpH can be positive, zero, or negative, depending on the net surface charge at the time of sampling and analysis of the soil (USDA, Soil Survey Staff, 1960). A positive value for ΔpH indicates the presence of negatively charged clay colloids. A negative value, on the other hand, means the presence of a positively charged clay colloid. The ZPC is reached when ΔpH equals zero. This is in general true, if only KCl is present in the system, since KCl is considered to be an indifferent electrolyte. If, however, other anions are present, which are subject to specific adsorption, they may shift the ZPC value to lower pH regions and consequently render the colloid surface more negative in charge (Mekaru and Uehara, 1972).

4.10 SURFACE POTENTIAL

From the above it is clear that if the H^+ and OH^- are the potential determining ions of reversible interfaces, adsorption of protons produces positive charge surfaces, whereas adsorption of OH^- yields negative charges. These reactions can be summarized in the following relationship (Gast, 1977; Van Raij and Peech, 1972):

$$-Al\text{-}OH^{0.5-} + H^+ \rightleftharpoons -Al\text{-}OH_2^{0.5+}$$

This relationship is of course dependent on pH, while at pH_0 (or ZPC) in order to maintain electroneutrality, the amount of positive charges must equal that of negative charges.

Because of the presence of opposite charges on the colloid surface and in the liquid phase, an electric potential develops at the solid-liquid interphase, called the *surface potential* ψ. The magnitude of the surface potential is given by the Nernst equation:

$$\psi = \frac{RT}{nF} \ln \frac{(H^+)}{(H^+)_{ZPC}}$$

Changing into common log gives

$$\psi = 0.059 \, [\log (H^+) - \log (H^+)_{ZPC}]$$

or

$$\psi = 0.059\ (ZPC - pH)\ \text{Volts at } 25°C$$

4.11 ELECTRIC DOUBLE LAYER

Because of the presence of an electronegative charge, clay in suspension can attract cations. These positively charged ions are not distributed uniformly throughout the dispersion medium. They are held on or near the clay surface. Some are free to exchange with other cations. The negative charge of the clay surface is thus screened by an equivalent swarm of counter ions that are positive. The negative charge on the clay surface and the swarm of positive counter ions are called the *electric double layer*. The first layer of the double layer is formed by the charge on the surface of the clay. Technically the charge is a localized point charge. However, we customarily consider this charge to be distributed uniformly over the clay surface.

The second layer of the double layer is in the liquid layer adjacent to the clay surface. The positive counter ions in this zone are attracted to the clay surface, but at the same time they are free to distribute themselves evenly throughout the solution phase. The two processes will come to an equilibrium, and the resulting distribution zone is like the distribution of gas molecules in the earth's atmosphere.

Helmholtz Double-Layer Theory

This is perhaps one of the earliest theories. The negative charge on the colloid is considered to be evenly distributed over the surface (charge density σ). The total counter charge in the second layer is concentrated in a plane parallel to the surface at distance x. If the medium has a dielectric constant D, then the electrokinetic potential

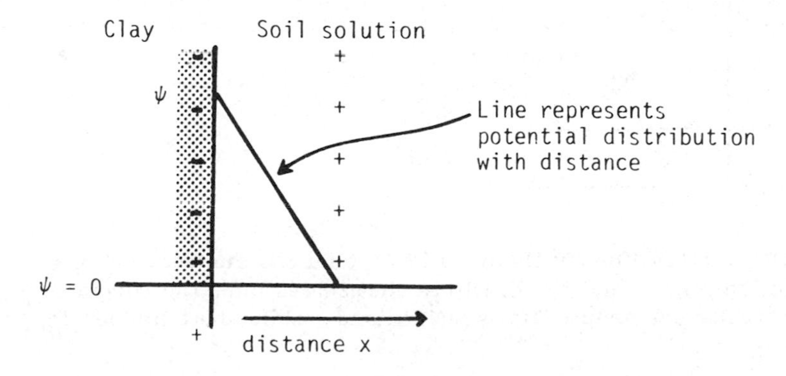

ζ is the same as the total potential ψ:

$$\psi = \frac{4\pi\sigma\, x}{D}$$

The electrochemical potential is maximum at the colloid surface and drops linearly at locations with increasing distance (x) from the surface within the double layer.

Gouy-Chapman Double-Layer Theory

The negative charge is again considered distributed evenly over the colloid surface. However, the counter ions are dispersed in the liquid layer as is the case of gas molecules in the earth atmosphere. This theory is also called the *diffuse double-layer theory* of Gouy and Chapman. The concentration distribution in the liquid zone follows the Boltzmann equation:

$$C_x = C^o_x \exp(-ze\psi/kT)$$

where

C_x = concentration of cations at distance x from surface
C^o_x = concentration of cations in the bulk solution
z = valence
e = electronic charge
ψ = electrical potential
k = Boltzmann constant
T = absolute temperature

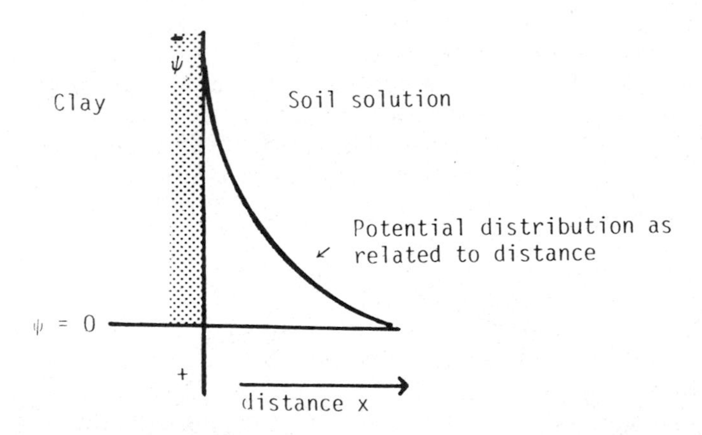

Because of the attraction by the negatively charged surface, cations in the solution phase tend to distribute themselves over the colloid surface so that electroneutrality is maintained, while the tendency for these ions to diffuse away is counteracted by Van der Waals attraction. A deficit of anions is usually present in the liquid interface, and the total charge of the surface is considered to be balanced by excess cations. The initial electric potential at the colloidal surface is maximum and decreases exponentially with distance from the surface as follows:

$$\psi_{\chi} = \psi_{o} \exp(-K_{\chi})$$

where

ψ_{χ} = electric potential at distance χ
ψ_{o} = surface potential
K = constant associated with concentration, valence of ions, dielectric constant, and temperature

At room temperature,

$$K = 3 \times 10^{7} z^{\pm} \sqrt{C}$$

where

z = valence of the ion
C = concentration of the bulk solution in moles per liter

The value 1/K is usually used as a measure of the thickness of the double layer (Verwey and Overbeek, 1948). As indicated by the formula for K, the thickness of the double layer is suppressed by both z and C. If C increases with a factor 4x, K increases with a factor of $\sqrt{4} = 2$. This means that the thickness of the double layer (1/K) is decreased $\frac{1}{2}$x from the surface. A similar discussion can be given for the valence z.

The Gouy-Chapman diffuse double layer theory is equally valid for positive charged colloidal surfaces. For a positive charged surface, excess anions will be present in the liquid interface, and a deficit of cations is then expected to occur at the surface.

Limitations to the Gouy-Chapman Diffuse Double-Layer Theory

The diffuse double layer theory was developed independently by Gouy (1910) and Chapman (1913) for the application on flat surfaces, but it may apply equally well to rounded or spherical surfaces (Verwey and Overbeek, 1948). The negative charge was considered to be evenly distributed over the surface. Since the counter ions are assumed to be point charges, and therefore occupy no spaces, they may reach excessively high concentrations at the liquid interface. Modification to the Gouy-Chapman theory was later presented by Stern (1924), who stated that ions of finite sizes can not approach the colloidal surface more closely than allowed by their effective radii.

Stern Double-Layer Theory

Stern made corrections in the double layer theory by taking into consideration the ionic dimensions. The influence of ionic dimension is greatest near the colloid. In the Stern theory, the first layer is similar to that of the previous theories. However, the second layer is divided into (1) a sublayer nearest the colloid surface and (2) a diffuse layer:

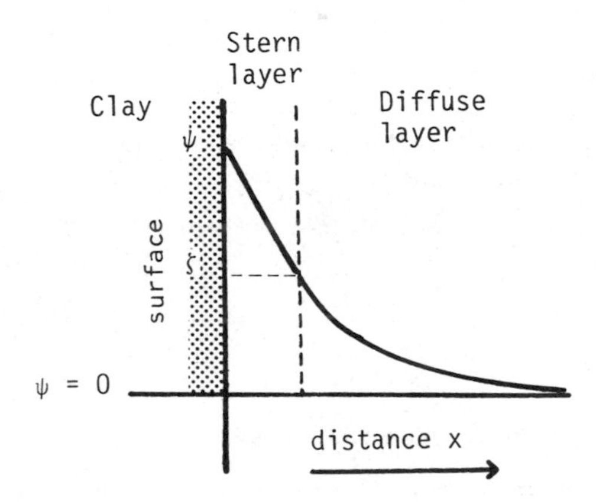

The first sublayer is tightly packed with cations, and is called the *Stern layer.* The potential distribution appears to be a combination of the Helmholtz and the Gouy-Chapman diffuse double layer. The decrease in potential is also divided into two parts. In the Stern layer, the potential decreases with distance from the surface according to the Helmholtz theory. From here on (in the diffuse layer), the decrease in potential with distance follows the Gouy-Chapman theory.

Effect of Electrolytes on the Thickness of the Diffuse Double Layer

The thickness of the diffuse double layer is dependent on the electrolyte concentration of the bulk solution (Table 4.13). High concentrations of electrolyte will usually result in suppression of the double layer. By increasing the electrolyte concentration, the amount of cations is increased. The latter reduces the concentration gradient in the liquid interface between the colloidal surface and surrounding liquid phase. Therefore, the tendency of cations to diffuse away from the surface of the colloids decreases, bringing about a decrease in thickness of the double layer.

Table 4.13 Effect of Concentration and Valences of Ions on Thickness of the Diffuse Double Layer

Electrolyte concentration, mol/l	Thickness of diffuse double layer (1/K), cm	
	Monovalent ions	Divalent ions
1×10^{-5}	1×10^{-5}	0.5×10^{-5}
1×10^{-3}	1×10^{-6}	0.5×10^{-6}
1×10^{-1}	1×10^{-7}	0.5×10^{-7}

Source: Verwey and Overbeek (1948).

Effect of Valency of Cations on the Thickness of the Diffuse Double Layer

The thickness of the diffuse double layer is also affected by the valency of the exchangeable cations. Generally, it has been reported that at equivalent electrolyte concentrations, monovalent cations in exchange position yield thicker diffuse double layers than divalent cations (Table 4.13). Trivalent cations will decrease the thickness of the double layer more strongly than divalent ions. This phenomenon is due to the tendency of ions to diffuse away from the colloidal surface (dissociate) being in the following decreasing order:

Monovalent ions > divalent ions > trivalent ions

For example, Na and K ions are frequently reported to be responsible for relatively thicker doubler layers than Ca and Mg, whereas double layers formed by Al ions are comparatively the thinnest.

The Zeta (ζ) Potential

When a colloidal suspension is placed in an electric field, the colloidal particles move in one direction (toward positive pole). The counter ions move into another direction (toward negative pole). The electric potential developed at the solid-liquid interface is called the *zeta* (ζ) *potential*. The seat of the ζ potential is the shearing plane or slipping plane between the bulk liquid and an envelope of water moving with the particle. Since the position of the shearing plane is not known, the ζ potential represents the electric potential at an unknown distance from the colloidal surface. Van Olphen (1977) stated that the ζ potential is not equal to the surface potential. It is less than the electrochemical potential on the colloid. Perhaps, it is comparable with the Stern potential.

Effect of Electrolytes on ζ Potential

The thickness of the double layer affects the magnitude of the ζ potential. Increasing the electrolyte concentration in the solution usually results in decreasing the thickness of the double layer. Compression of the double layer will also occur by increasing the valence of the ions in the solution.

The ζ potential may, therefore, be expected to decrease with increasing electrolyte concentration. It reaches a critical value, at the point at which the ζ potential equals zero. This point is called the *isoelectric point*. At the isoelectric point, the double layer is very thin and particle repulsive forces are at a minimum. At and below this point repulsion would no longer be strong enough to prevent flocculation of colloidal particles. The ζ potential is not a unique property of the colloid, but depends on the surface potential (ψ) of the clay particle. It is determined from the electrophoretic mobility of the suspension using the formula

$$V_e = \frac{D \zeta E}{4 \pi \eta}$$

where

V_e = electrokinetic velocity
D = dielectric constant
E = applied emf
η = viscosity of the fluid

The ζ potential is in fact the electrokinetic potential at the slipping plane surface:

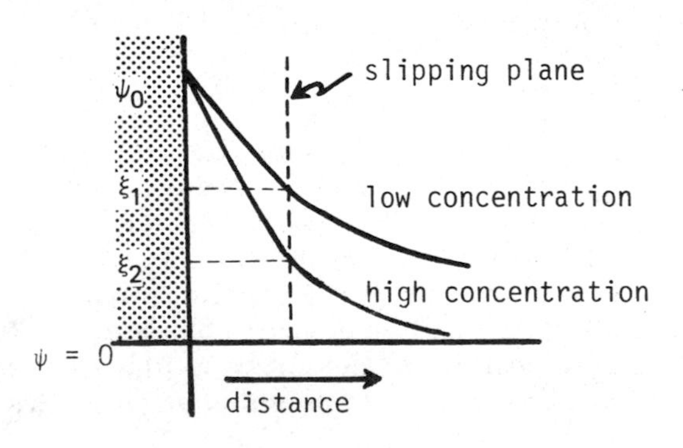

The surface potential of the colloid is ψ_0. In dilute solution, the electrokinetic potential has a value represented by ζ_1. By adding salt to the solution, the diffuse layer is suppressed and more counter ions are forced to the colloid surface within the slipping plane. The slipping plane is considered to be at a fixed distance. Hence, at high salt concentration a change occurs in the total potential distribution as related to distance from the colloid. The potential distribution at high salt concentration is represented by the bottom curve in the figure. The electrokinetic potential ζ_2 is, therefore, smaller than ζ_1.

4.12 THE ELECTRIC DOUBLE LAYER AND STABILITY OF CLAYS

As discussed previously, clays carry a negative charge, which is ordinarily balanced by exchangeable cations adsorbed on their surface. In suspension, the cations tend to diffuse away from the clay surface into the bulk solution to balance the concentration difference occurring between the interface and bulk liquid phase. However, a large portion of these ions, especially those in the immediate vicinity of the clay surface, can not move very far away, because of the strong attraction from the negative charge on the clay surface. The cations aggregate in the interface thus forming an electric double layer, which may vary in thickness from 50 to 300 Å.

Whenever such clay particles approach each other, repulsion between the particles occurs, because the outer parts of the double layers have the same type of charge (positive). The suspension is then considered stable, and the clay is considered to be dispersed. Because of this approach, the diffuse counter-ion atmospheres of the two particles interfere with each other. This leads to a rearrangement of the ion distribution in the double layers of both particles. Work must be performed to bring about these changes. The amount of work to bring about the changes is called *repulsive energy* or *repulsive potential* V_r at the given distance. The range and effectiveness of the repulsive potential depend on the thickness of the double layer. The repulsive force decreases usually exponentially with increasing distance between the particles (see Figure 4.46). Opposite to repulsive forces, the clay suspension is also subjected to interparticle attraction. These forces of attraction are usually called the *Van der Waals attraction* (V_A). Van der Waals attraction is only effective at very close distances, and decays rapidly with distance. However, since it is additive between atom pairs, the total attraction between particles containing a large number of atoms is equal to the sum of all attractive forces between every atom of one particle and every atom of the other particle.

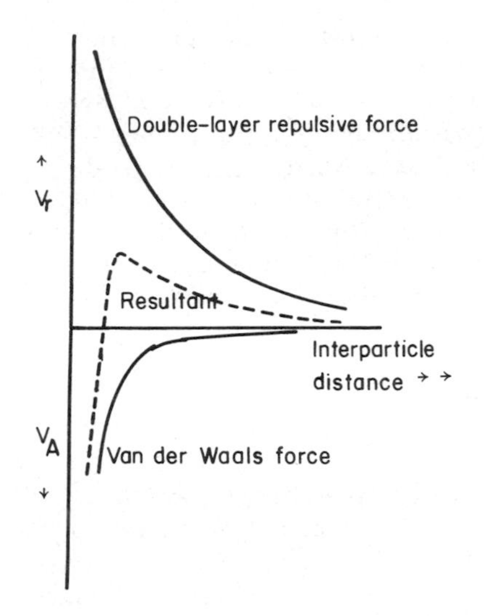

Figure 4.46 Double-layer repulsive and interparticle attractive (Van der Waals) forces as a function of interparticle distance.

When the interparticle distance decreases to about 20 Å or less, Van der Waals forces will become dominant, and the clay particles will flocculate. At interparticle distance of > 20 Å, repulsive forces are dominant, creating a stable clay suspension. An example of the resultant of repulsive and Van der Waals forces is shown in Figure 4.46 by the broken line curve.

Repulsion will dominate at low electrolyte concentration. The clay particles are shielded by relatively thick double layers, decreasing the possibility of mutual approach. At high electrolyte concentration, the chances of close approach are made possible by compression of the double layers. In this condition, Van der Waals attraction may overcome the repulsive forces and coagulation or flocculation of colloidal particles occurs rapidly.

4.13 THE EFFECT OF FLOCCULATION AND DISPERSION ON PLANT GROWTH

The problem of flocculation and stability of soil suspensions is very important in soils. Stable aggregates can only be formed in soils containing clay that will flocculate. If clays remain dispersed, the soil is puddled. Puddled soils are sticky when wet and hard when dry.

Root growth and soil aeration require a porous condition in soils. If percolating rainwater leaches out electrolytes from the soil, clay particles may become dispersed. As the soil becomes dry, caking or soil compaction may occur. The latter reduces the pore spaces, which inhibit soil aeration necessary for adequate root growth. Therefore, a flocculating concentration of electrolytes should be maintained in the soil. To reach such a condition, the soil should be limed, although acid soils high in Al are usually flocculated. Calcium and Mg are known to have high flocculation powers on the negative clay particles and will reduce the toxicity effect of high Al concentrations.

5

ADSORPTION IN SOILS

The electrochemical properties discussed in the preceding sections find many practical applications in soils. Besides the effect of flocc-ulation on soil condition and plant growth, there are reasons for the soils developing the capacity to adsorb gas, liquid, and solid con-stituents. Cation exchange reactions, interactions between clay and organic compounds, complex reactions between metal ions and inorganic and organic colloids are additional implications of the electrochemical behavior of soil colloids.

5.1 FORCES OF ADSORPTION

Forces responsible for adsorption reactions include (1) physical forces, (2) hydrogen bonding, (3) electrostatic bonding, and (4) coordination reactions.

Physical Forces

The most important is the Van der Waals force. This force is a re-sult of short-range dipole-dipole interactions. Its role is only of importance at close distances since this type of force decreases rapidly with distance.

Hydrogen Bonding

As defined earlier, the bond by which a hydrogen atom acts as the connecting linkage is called a *hydrogen bond*. Water, which is dipolar, may become adsorbed at the clay surface through its linkage with hydrogen bonding.

Electrostatic Bonding

Another type is electrostatic attraction of ions, which is the result of the electrical charge on the colloid surface. This is the reason for cation exchange reactions on clay surfaces.

Coordination Reactions

The reaction involves coordinate covalent bonding. The latter occurs when the ligand donates electron pairs to a metal ion. The compound formed is called a *complex compound* or a *metal complex*. In cases where they involve a reaction between an organic ligand and a metal ion only, the distinction between adsorbate and adsorbens may become obscure.

5.2 ADSORPTION

Adsorption was defined earlier as the concentration of constituents at the colloidal surfaces. The curve relating the concentrations of materials adsorbed at a fixed temperature is called the *adsorption isotherm*. Four major types of equations are used to describe adsorption isotherms: (1) the Freundlich equation, (2) the Langmuir equation, (3) the BET (Brunauer, Emmitt, and Teller) equation, and (4) the Gibbs equation.

Freundlich Equation

The adsorption isotherm in many dilute solutions is formulated by Freundlich (1926) as

$$\frac{x}{m} = kC^{1/n} \tag{5.1}$$

where

x = amount of material adsorbed
m = amount of adsorbens
C = concentration of the equilibrium solution
k, n = constants

The value of $1/n$ is usually between 0.2 and 0.7 (Kruyt, 1944). For many pesticides at dilute concentrations $1/n = 1$ (R.A. Leonard, personal communication.) The equation has no theoretical foundation, and is empirical in nature. The curve according to Eq. (5.1) is usually parabolic. However, by taking the logarithm, Eq. (5.1) changes into

$$\log \frac{x}{m} = \log k + \frac{1}{n} \log C$$

The log equation gives a "straight-line" curve.

Langmuir Equation

Another method to express adsorption is given by Langmuir (1916-1918):

$$\frac{x}{m} = \frac{k_1 C}{1 + k_2 C} \tag{5.2}$$

where

x = amount adsorbed
m = amount of adsorbens
k_1, k_2 = are constants
C = concentration of equilibrium solution

The difference with the Freundlich equation is the following. At very high concentration k_2C in Eq. (5.2) reaches such a value that the factor 1 can be neglected, so that the formula changes into

$$\frac{x}{m} = \frac{k_1}{k_2}$$

The latter formula stated that x/m becomes constant at high concentration. In other words, at high values of C, the surface of the adsorbens becomes saturated, and adsorption reaches a maximum.

Brunauer, Emmett, and Teller (BET) Equation

This equation was developed by the above authors in 1938 for the adsorption of multilayers of nonpolar gases. The equation at low pressure is as follows:

$$\frac{P}{V(P_o - P)} = \frac{1}{V_m C} + \frac{C - 1}{V_m C}\frac{P}{P_o} \tag{5.3}$$

where

P = equilibrium vapor pressure
P_o = saturation vapor pressure
V = volume of gas adsorbed
V_m = volume of gas adsorbed when solid is covered with monolayer
C = constant related to heat of adsorption

The BET equation is an extension of the Langmuir for application to multilayer adsorption. It is assumed that the first layer of gas is attracted firmly to the surface, perhaps by Van der Waals forces. The second and subsequent layers are held by weaker forces. As P/P_o increases, the layers of gas are building up in an unrestricted way. The number of layers becomes infinite when P/P_o is unity. If $P/[V(P_o - P)]$ is plotted against P/P_o (as the abscissa), a straight line curve should be obtained. The slope is characterized by the factor $(C - 1)/V_mC$, and the intercept is at $1/V_mC$.

Gibbs Equation

This equation describes adsorption processes in relation to surface tension:

$$\Gamma = -\frac{a}{RT}\left(\frac{\partial \gamma}{\partial a}\right)_T \tag{5.4}$$

where

Γ = surface concentration of adsorbed material
a = activity of solute in moles
R = gas constant
T = absolute temperature
γ = surface tension in dyn/cm

The solute is adsorbed on the surface of the adsorbens if $\partial \gamma/\partial a$ is negative. If $\partial \gamma/\partial a$ is positive, the solute is more concentrated in the bulk solution than in the interface region. The latter is sometimes called *negative adsorption*. The Gibbs equation finds primary application in adsorption at liquid-gas interfaces.

5.3 ADSORPTION OF WATER

Water is held in the soil in the pore spaces by forces of attraction at the colloid surfaces, by surface tension in the capillaries, and by attraction to the ions. In a wet condition, all the pores of the soil are filled with water. The soil is considered to be at its "maximum retentive capacity." In this condition the matric potential ψ_m is 0. The excess water is free to move in the soil by gravity. The movement, called *drainage*, usually results in a number of undesirable effects, such as waterlogging and leaching of nutrients.

As soon as excess water has drained away by gravity, water occurs in the macropores as thin layers (films) on the surfaces of soil particles, and as wedges at the points of contacts between the particles. The micropores, on the other hand, are still filled with water. Such a condition is called *field capacity*. The force holding this water is approximately 0.2 to 0.3 bars. Living plants must satisfy their water need from field capacity water, by applying suction forces that can overcome the forces of adhesion, cohesion, and osmosis. As the water is taken up by the roots, the water films on soil particle surfaces become thinner. Hence, the forces of adsorption and retention become larger and larger. A point will eventually be reached where the attraction of water by soil particles becomes so large, that the plants can no longer extract enough to satisfy their needs. This point is called the *permanent wilting point*. The amount of water at this point is called the *permanent wilting percentage*. The force holding this water to the soil particles is estimated to be 15 bars.

Adsorption of Water by Silicate Clays

The adsorption of water by silicate clays is attributed to electrical forces. Upon adsorption, orientation of water takes place by the electrical field on the surface of clays, and the water molecules loose some of their freedom of movement. In terms of thermodynamics, it is said that the free energy of water has been decreased upon adsorption. This part of the energy is lost in the form of heat as noticed by an increase in temperature at the wetting front as water reacts with soil. According to the BET adsorption theory, the heat of adsorption of the first layer of water is given by the constant C in Eq. (5.3). C itself can be found by the following relation:

$$C = a \exp \frac{Q_1 - Q_L}{RT}$$

where

a = proportionality constant
Q_1 = heat of the first adsorbed layer of water
Q_L = heat of liquefaction (becoming liquid)
R = gas constant
T = absolute temperature

For practical purposes, $a = 1$ (Taylor and Ashcroft, 1972) and the equation can be changed to

$$2.303 \log C = \frac{Q_1 - Q_L}{RT}$$

The heat of adsorption, or the difference in energy of adsorption at the interface, $Q_1 - Q_L$, varies with the type of clay minerals. Taylor and Ashcroft (1972) reported that it was low for illites and other 2:1 types of minerals, but that it was comparitively higher for kaolinite and other 1:1 types of clays.

Clays with large internal surfaces are also capable of adsorbing water in the intermicellar regions, and interlayer hydrates can be formed. The rate of stability and the structure and characteristics of interlayer water are dependent on the presence of interlayer cations and on the composition of the interlayer clay surfaces. Interlayer water reacts both with the oxygens of the siloxane surface and with cations present in the intermicellar spaces. In the event that only one to three layers of water are adsorbed, some degree of water stability has been noted. Because of the strong bonding of this water, the behavior of the adsorbed interlayer water is relatively static. On the other hand, when more water layers are adsorbed in internal spaces, this water behaves more liquidlike, and a higher degree of mobility is then exhibited. Interlayer water can form a link between cations and the negative charged clay surface. They are also capable of donating protons to bases in interlayer positions, and can, as well, donate or accept electrons from two neighboring ions. Proton transfer from water to bases often takes place more readily in the intermicellar spaces than in the free soil solution. Formation of protonated bases in interlayer solutions is considered the result of increased acidity and ionization of interlayer water.

The adsorption of water in intermicellar spaces of expanding 2:1 lattice type of clays, e.g., montmorillonite, usually results in an increase in basal spacings, as determined by x-ray diffraction analysis (Table 5.1). The data in Table 5.1 seem to confirm that amount of water present and nature of interlayer cation both have an influence on adsorption. However, increased humidity appears to expand the lattice space more than the type of cation.

Two types of interlayer water can be distinguished (Walker, 1961): (1) water in the first (inner) hydration shell coordinated around the interlayer cation and (2) water in the secondary (outer) coordination sphere of the cation. The latter is more mobile than water in the first coordination sphere. The presence of these two types of interlayer water has been discovered in x-ray analysis and infrared spectroscopy. With infrared analysis water in the two coordination spheres is distinguished by differences in intensity of infrared absorption at 3600 to 3400 and at 1620 cm^{-1}, for OH-stretching and OH-bending vibrations, respectively, of the water molecule. Water in the first coordination sphere has an infrared absorption at 3600 cm^{-1} and a very strong absorption at 1620 cm^{-1}, perhaps because of the weaker hydrogen bonding involved in the adsorption process. In contrast, water in the second coordination sphere, or the more mobile water, exhibits a spectrum similar to that of free water, with absorption bands between 3600 and 3400 cm^{-1} and a relatively weak band at 1620 cm^{-1}. The OH-stretching vibration of interlayer water between 3600 and 3400 cm^{-1} is usually difficult to observe, since structural OH groups of montmorillonite may also absorb at or near 3600 cm^{-1}.

Interlayer water can be displaced by other polar compounds. Generally water adsorbed in the second coordination sphere is more easily displaceable than water held in the first coordination shell. A number of polar compounds, such as pyridine and nitrobenzene, are capable of displacing the mobile interlayer water of the outer shell. Removal of water from the first coordination shell takes place only by treatments which effect dehydration of the cation. This displacement of water from the primary shell causes the cation to polarize the residual water with the consequent dissociation of the water molecules. The protons produced by the dissociation reaction may be used for protonation of other organic compounds.

Table 5.1 Basal, d (001) Spacings in Å, of Montmorillonite Saturated with Different Cations, as Affected by Interlayering with Adsorbed Moisture

	Interlayer cations					
	Li^+	Na^+	K^+	Ca^{2+}	Mg^{2+}	Ba^{2+}
Increasing humidity ↓	9.5	9.5	10.0	9.5	9.5	9.8
	15.4	15.4	15.0	15.4	15.4	15.5
	19.0	19.0	—	18.9	19.2	18.9
	22.0	—	—	—	—	—

Source: Norrish (1954).

Water Adsorbed by Organic Matter

Organic matter is known to contain a number of functional groups, such as carboxylic groups, phenolic and alcoholic hydroxyl groups, amino acid groups, amides, ketones, and aldehydes. Among the many groups above, perhaps the most important sites for adsorption of water are provided by the carboxyl groups. Ionized carboxylic groups possess high affinity for water, although the other functional groups may also exhibit some degree of adsorption.

Water can be held by a single hydrogen bond or by multiple bonds or coordination bonds. Coordinated water is usually held more strongly by organic matter than is water held by a single bond. Upon drying and exposure of polar sites of organic matter, an internal pairing of proton donor groups (OH) and proton acceptor groups (C=O) may occur. These reactions can produce stable sites of pairing, so that the organic compound will not rehydrate upon wetting. The latter is possibly part of the reasons for the irreversible behavior of organic matter in rehydration upon wetting and drying.

5.4 PLANT–SOIL-WATER ENERGY RELATION

As discussed previously the water potential of plant cells is defined as

$$\psi_{w(plant)} = \psi_p + \psi_s + \psi_m$$

When a plant root is immersed in the soil solution, water moves into or out of the plant cells, until equilibrium is reached. At equilibrium the water potential of the plant cells equals the water potential of the soil solution:

$$\psi_{w(plant)} = \psi_{w(soil)}$$

Substituting this relation in the above equation gives

$$\psi_{w(soil)} = \psi_p + \psi_o$$

where $\psi_o = \psi_s + \psi_m$ = osmotic potential

By rearranging the equation we find

$$\psi_p = \psi_{w(soil)} - \psi_o \quad (5.5)$$

The permanent wilting point occurs at $\psi_p = 0$ (Slatyer, 1957). Therefore

$$\psi_{w(soil)} - \psi_o = 0$$

or

$$\psi_{w(soil)} = \psi_o$$

If the soil-water potential decreases, soil-water content also decreases. The latter may cause water to move out of the cell, ultimately producing a phenomenon called *cytoplasmolysis*.

5.5 ADSORPTION OF ORGANIC COMPOUNDS

Adsorption of organic compounds by soil components has recently received increasing attention, especially adsorption of pesticides and herbicides. Physical, hydrogen, electrostatic, and coordination bonding are involved in the process (Bailey and White, 1970). Physical bonding, due to the London—Van der Waals attraction, is limited to external surface reactions only. Hydrogen bonding may occur if the compound has an N—H or OH group that can be linked to the O on the clay surface. Electrostatic bonding takes place between organic cations and anions and clays. The organic cations are mostly adsorbed on the negative clay surfaces, whereas the organic anions are more attracted towards the edge surfaces of clay (Van Olphen, 1977). Intermicellar adsorption between the clay layers is limited to low molecular weight organic compounds (Tan and McCreery, 1975). The presence of organic compounds in intermicellar positions results in an expansion of the basal spacings. For example, montmorillonite treated with ethylene glycol or glycerol, gives a basal spacing of 17 Å, in contrast to 9.2 to 9.5 Å, its ordinary spacing in the presence of water and/or exchangeable cations. In the adsorption process, the organic molecule may displace adsorbed water from the clays. In turn, the adsorbed organic compound can often be removed by washing with water, except perhaps for those in the intermicellar spaces. Coordination bonding results in formation of complex compounds. It is formed by the organic ligand donating electron pairs to a metal or an ion. The metal takes the central position and is surrounded by a group of ligands. The metal ion, complexing organic matter, can also be attached to a clay particle. The metal then forms a bridge between the organic compound and the clay surface.

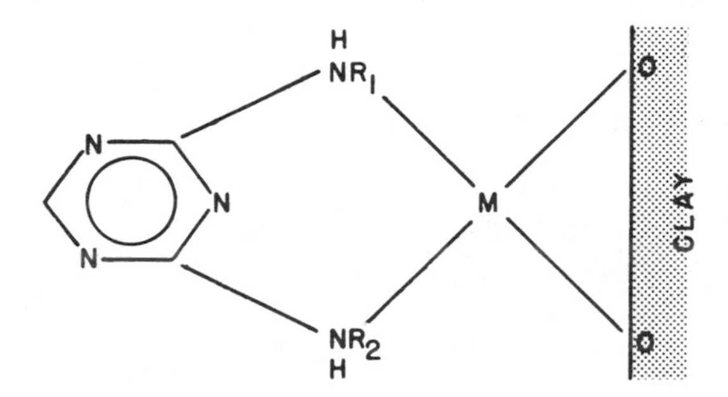

Uncharged organic compounds are currently also known to be subject to adsorption by clay minerals. The adsorption can be direct or indirect, and physical forces, e.g., Van der Waals forces, are mainly responsible for such an attraction. In order for the physical force to become effective, the uncharged organic compounds have to be relatively large in size or chain length. A chain length of maximum five units has been reported to effect adsorption significantly (Theng, 1974). Molecules of larger size, or with more than five units, can be adsorbed in the presence of excess water. The increase in adsorption with longer chain length is attributed to Van der Waals forces becoming more effective as size of molecules is increased.

Indirectly, uncharged organic compounds are adsorbed by the silicate surface through linkage to the exchangeable cations. Evidence for the latter came from infrared spectroscopy. Competition for ligand sites usually occurs between the organic compounds and water molecules around the exchangeable cations. The organic compound can react either directly with the cation, or indirectly be coordinated to the cation by linkage with water molecules in the hydration shell. This type of reaction, called *ion-dipole reaction*, depends on the polarizing power of the cation, the basicity of the organic compound, and on the nature of packing when adsorbed in intermicellar spaces.

Another adsorption mechanism reported in the literature is the adsorption caused by the presence of activated C—H groups. Activation of methylene groups by adjacent electron acceptors makes possible for the activated methylene groups to form hydrogen bonds with the oxygens of the siloxane surface of clay minerals. However, in later studies many arguments have been presented on the possibility for formation of a C—H···O type of bonding.

Nature of Adsorption Isotherms

No general agreement exists in the literature as to the nature of adsorption isotherms. Bailey and White (1970) are of the opinion that adsorption of organic matter can be better described with the Freundlich equation, but Weber (1970) has reported that adsorption of organic matter follows the Langmuir equation. According to the Freundlich equation, theoretically, adsorption increases indefinitely with increasing concentration. On the other hand, the Langmuir equation indicates that adsorption of organic matter on the clay surface tends to reach a maximum limit. The latter is compatible with the fact that soil and clay do not have an infinite capacity to adsorb, but will sooner or later be saturated. Adsorption of organic compounds, such as humic acid and the like, following the Langmuir type of equation has also been reported by Inoue and Wada (1973) and Tan et al. (1975). In their studies with poultry litter extracts adsorbed by kaolinite and montmorillonite minerals, Tan et al. (1975) show that the slope of the adsorption isotherm decreases with increasing temperature from 25 to 35 to 50°C (see Figure 5.1). They also indicate that infrared analysis confirms that protonated compounds are adsorbed in larger amounts than Na^+-saturated extract. The latter may suggest that ionic bonding has been involved in the adsorption reaction.

Effect of Molecular Size on Adsorption

The size of the organic molecule is considered to play an important role in its rate of adsorption. Bailey and White (1970) summarize the effect of molecular size as follows:

1. Adsorption of nonelectrolytes by nonpolar adsorbents increases as molecular weights of the substances increases.
2. Van der Waals forces of adsorption increases with increasing molecular size.
3. Adsorption decreases because of steric hindrance.

The evidence currently available also shows the presence of a maximum limit in molecular size in adsorption of organic compounds. As stated earlier, appreciable adsorption is noticed with compounds having a chain length of five units, while larger molecules (chain length greater than five units) may be adsorbed only in the presence of excess water. However, very large molecules can experience difficulties in adsorption due to adverse molecular configuration. Inoue and Wada (1973) succeeded in determining the molecular size limit for adsorption. They reported that humic molecules with molecular weights between 1500 and 10,000 are preferentially adsorbed over the smaller or larger molecules. These size limits have

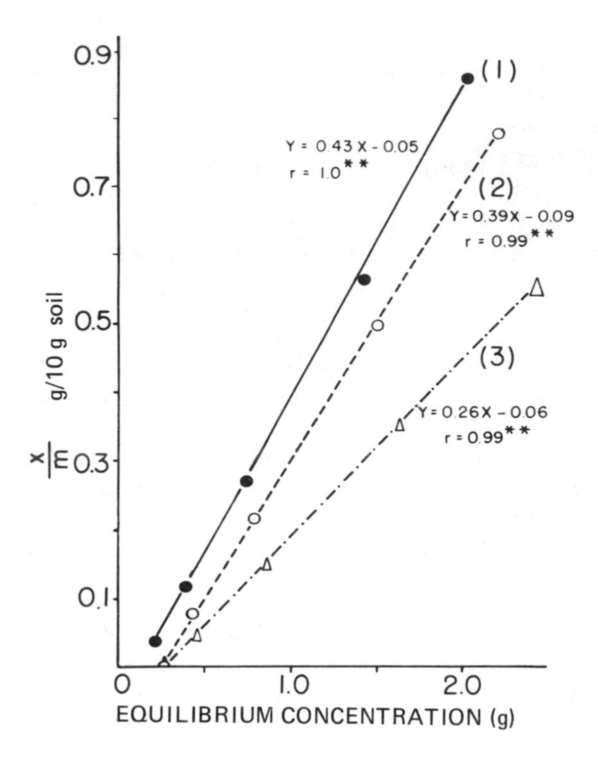

Figure 5.1 Adsorption of water extracts of broiler litter by Cecil topsoil at (1) 25°C, (2) 35°C, and (3) 50°C. [Adapted with permission from Tan et al. (1975). Copyright 1975 American Chemical Society.]

been confirmed by Tan (1976a) in his studies with Sephadex gel filtration. To study complex formation, humic acid was shaken with kaolinite or montmorillonite, and the remaining humic acid in solution was fractionated with Sephadex G-50 gel filtration. The results (see Figure 5.2) showed that mainly the fraction with a molecular weight of 1500 (HA-II) was adsorbed. Prior to treatment with kaolinite or montmorillonite, the original humic acid was composed of equal amounts of high (HA-I: >30,000) and low-molecular-weight (HA-II: <1,500) fractions. After interaction with the clay minerals, only the elution peak of the low-molecular-weight fraction was decreased in intensity, meaning that the equilibrium humic acid solution has lost some of the low molecular weight humic acid. The amounts of humic acid lost had apparently been adsorbed by kaolinite or montmorillonite. The elution curves also provide indications that montmorillonite has a larger adsorption capacity than kaolinite. The elution peak of HA-II was significantly smaller after treatment with montmorillonite than with kaolinite. This difference in adsorption due to type of clays has been expected. The expanding 2:1 minerals have properties, such as high CEC and surface area, that give rise to large coulombic and Van der Waals forces. On the other hand, the nonexpanding 1:1 type of clays have low CEC and low surface area, all properties that will not contribute to considerable adsorption.

Interlayer Adsorption and Molecular Orientation of Organic Compounds

Adsorption of organic compounds by clay surfaces can occur not only on the outer surfaces, but x-ray diffraction analysis of basal spacings of clays has shown that considerable amounts of the organic molecules can penetrate the intermicellar regions of 2:1 expanding types of

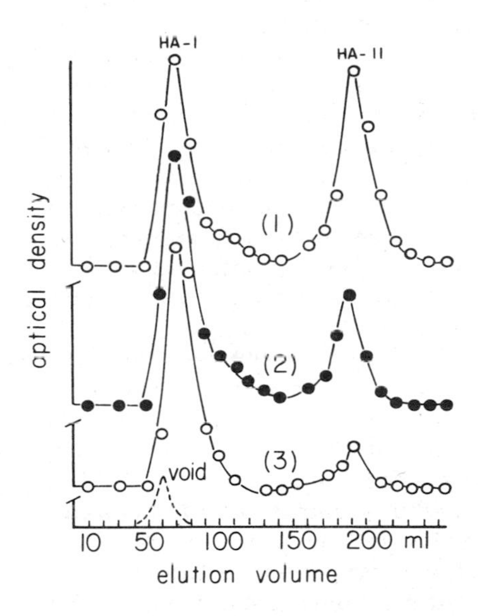

Figure 5.2 HA-I is high-molecular-weight humic acid (MW > 30,000) and HA-II is low-molecular-weight humic acid fractions (MW < 1,500) separated by Sephadex G-50 gel filtration: (1) original humic acid (2) humic acid after interaction with kaolinite, and (3) humic acid after interaction with montmorillonite. (Reprinted with permission from Soil Biol. Biochem., 8: 235-239, K.H. Tan, "Complex formation between humic acid and clays as revealed by gel filtration and infrared spectroscopy. Copyright 1976 Pergamon Press, Ltd.)

clays especially. This placement of organic compounds in interlayer spaces of silicate minerals is called *intercalation* or *solvation*. Intercalation occurs more with the 2:1 than with 1:1 types of minerals. Because of the strong bonds between the layers of kaolinite, penetration of intermicellar spaces by organic compounds is very difficult, although intercalation can in some instances be achieved with certain inorganic salts and polar substances. Adsorption of organic compounds are, therefore, confined in general to the outer surfaces and to the edges of kaolinite, where the unsatisfied valencies by broken bonds have been estimated to amount to approximately 10 to 20% of the crystal area. In contrast to kaolinite, montmorillonite has both active outer and interlayer surfaces. Of the latter two, the interlayer surface accounts for the major portion of the total surface area. Only 10% of the total surface area is made up of active crystal edges, reducing in this way the importance of edge effects in adsorption.

Interlayer adsorption is affected by molecular size, polarity, and polarizability of the organic substances. In addition, it appears from previous discussions that water molecules and inorganic cations in intermicellar positions may also play an important role in adsorption. In order for the substance to be adsorbed on the intermicellar surfaces, sufficient energy must be available to exceed, or at least equal, the energy forces holding the layers together. The presence of polar molecules, such as water, may serve as an aid to the latter, since interlayer water is capable of keeping the layers apart, and in this way decreasing the electrical field between the clay surface-exchangeable ion. Evidence for the fact above was obtained with benzene, which can be adsorbed by montmorillonite only when interlayer water is present (Bailey and White, 1970).

Polarizability is the ease with which negative and positive charges in a molecule can be displaced with respect to one another in the presence of an electric field. From the definition it is perhaps clear that the greater the ease of polarizability of the molecule, the greater will be its adsorption in intermicellar spaces.

X-ray diffraction analysis of expanding silicate clays solvated with ethylene glycol reveals that polarity and polarizability play an important role in orientation and spatial arrangement of interlayer organic molecules. Two kinds of orientation have been reported for aliphatic chain molecules adsorbed with their axis perpendicular to the silicate surface (Theng, 1974): (1) an interlayer arrangement in which the plane of carbon zig-zag is perpendicular to the silicate layer, and (2) an interlayer arrangement in which the zig-zag carbon plane is parallel to the silicate surface. Strongly polar compounds will favor orientation according to arrangement (2), whereas nonpolar compounds tend to arrange themselves according to orientation type (1).

5.6 INTERPARTICLE ATTRACTION

Interaction between charged colloidal particles occurs frequently in soils. As discussed previously the presence of a double layer creates repulsive forces between particles. The colloids also possess London-Van der Waals forces responsible for attraction of particles to each other. The latter can be significant at close range, but will be of no importance at large interparticle distances. The net balance between the repulsive and attractive forces determines interparticle attraction. For most soils; Al^{3+}, H^+ or Ca^{2+}, and Mg^{2+} ions are dominant in the double layers. These ions, except H^+, tend to reduce the repulsive forces of the double layer. The Van der Waals forces will then be effective and the soil particles interact and form aggregates. In contrast to the ions above, Na^+ ions have a different effect. Due to its large hydration shell, Na^+ ions tend to increase the double-layer repulsive forces. The soil particles are not aggregated and when wet remain in a dispersed condition. Soils in which Na^+ ions are dominant, such as the saline soils, may have a poor physical condition.

6

CATION EXCHANGE

6.1 ADSORPTION OF CATIONS BY SOIL COLLOIDS

Since clay colloids carry negative charges, cations are attracted to the clay particles. They are held electrostatically on the surface of the clay. Most of them are free to distribute themselves through the liquid phase by diffusion. The density of ion population is, of course, greatest at or near the surface. These cations are called *adsorbed cations*. Different orders of adsorption are known among the cations. Generally ions with the smaller hydrated size are preferably adsorbed. The following decreasing order of preference for adsorption of monovalent cations by clays has been reported (Gast, 1977):

$$Cs > Rb > K > Na > Li$$

Such a series of ions of decreasing preferential adsorption is called a *lyotropic series*. Lyotropic series for polyvalent cations has also been mentioned in the literature (Taylor and Ashcroft, 1972). Evidence is present that different lyotropic series exist for different types of clays.

6.2 CATION EXCHANGE REACTIONS

The term *cation exchange* is preferred over the term *base exchange* since the reaction also involves H^+ ions. The hydrogen ion is a cation, but not a base. The adsorbed cations can be exchanged by other cations. The process of replacement is called *cation exchange*. The rate of reaction is virtually instantaneous. To maintain electroneutrality in the soil, exchange reactions are stoichiometric reactions as illustrated by the classical experiment of Way (1850):

$$Ca - soil + 2NH_4^+ \rightarrow (NH_4)_2 - soil + Ca^{2+}$$

Adsorption and cation exchange are of great practical significance in nutrient uptake by plants, soil fertility, nutrient retention, and fertilizer application. Adsorbed cations are generally available to plants by exchange with H^+ ions generated by the respiration of plant roots. Nutrients added to the soil, in the form of fertilizers, will be retained by the colloidal surfaces and are temporarily prevented from leaching. Cations that may pollute the groundwater may be filtered by the adsorptive action of the soil colloids. As such, the adsorption complex is considered to give to the soil a storage and buffering capacity for cations. In addition, it may play a role in making liming materials available to plant growth. Calcitic limestone, or $CaCO_3$, is insoluble in water. When added to acid soil (Al-soil), limestone may react with H_2O containing CO_2:

$$CaCO_3 + H_2CO_3 \rightarrow Ca(HCO_3)_2$$

The calcium bicarbonate formed is soluble in water. Ca^{2+} that is dissociated off can then be adsorbed by the soil in exchange for Al^{3+}:

$$\frac{3}{2}\,Ca(HCO_3)_2 + Al\text{-soil} \rightleftharpoons (Ca)_{\frac{3}{2}} - soil + Al(OH)_3 + 3CO_2$$

Thomas (1974) considered the processes above a neutralization and precipitation type of cation exchange reaction.

6.3 CATION EXCHANGE CAPACITY

The *cation exchange capacity* (CEC) of soils is defined as the capacity of soils to adsorb and exchange cations. It is usually expressed in milliequivalents per 100 grams of soils. However, sometimes the USDA Soil Survey Division uses as a unit milliequivalents per 100 grams of clay.* It is common practice in the determination of CEC to analyze all exchangeable cations. The CEC is then

$$CEC = \Sigma \text{ mEq exchangeable cations per 100 g soil}$$

Bolt et al. (1976) are of the opinion that a certain correction is needed to the above procedure. They stated that the real adsorbed cations are not accompanied by anions. But "free" cations may have slipped in carrying with them their counter anions. They may have been analyzed together with the exchangeable cations. The ions of these free salts must be subtracted to obtain the correct CEC.

---

*The unit used by the International System (SI) to express mEq per 100 g is cmol (p^+) kg^{-1}.

Example: Determination of exchangeable cations and anions gives the following results:

Cations, mEq per 100 g		Anions, mEq per 100 g	
Na^+	= 5	Cl^-	= 0.8
K^+	= 5	HCO_3^-	= 0.2
Ca^{2+}	= 10	Σ anions	= 1.0
Mg^{2+}	= 10		
H^+	= 5		
Σ cations	= 35		

The CEC, according to Bolt et al. (1976) is

$$\begin{aligned} CEC &= \Sigma \text{ exchangeable cations} - \Sigma \text{ exchangeable anions} \\ &= 35 - 1.0 = 34 \text{ mEq per 100 g} \end{aligned}$$

The value for soil CEC varies according to type and amounts of colloids present in soils. On the average the CEC of the major soil colloids is as follows:

Soil colloids	CEC, mEq per 100 g
Humus	200
Vermiculite	100-150
Montmorillonite	70-95
Illite	10-40
Kaolinite	3-15
Sesquioxides	2-4

6.4 THE EXCHANGING POWERS OF CATIONS

Different cations may have different abilities to exchange adsorbed cations. The amount adsorbed is often not equivalent to the amount exchanged. Divalent ions are usually held more strongly than monovalent ions. They will be exchanged with more difficulty. It is sometimes noticed that if Ba^{2+} is used as the exchange cation, the exchange does not occur in equivalent amounts. Barium is strongly adsorbed by the clay, but appears to have low penetrating power.

Therefore, it exchanges less than would be expected from the amount of Ba adsorbed. On the other hand, cation exchange using NH_4^+ ions may often yield higher exchange results than is expected from amounts of NH_4^+ ions adsorbed. Ammonium as a monovalent ion will be attracted less strongly than the Ba ions, but NH_4^+ has high penetrating power.

An exception to the above is perhaps the use of H^+ ions. Hydrogen ions are adsorbed more strongly than the other monovalent or divalent ions. Hydrogen clays prepared by exchange reactions contain initially large amounts of exchangeable H^+ ions and small amounts of Al^{3+} ions. However, the concentration of exchangeable Al^{3+} ions builds up rapidly. The exchangeable H^+ ions cause a partial decomposition and Al released from the clay becomes exchangeable.

6.5 THE IONIC COMPOSITION OF THE EXCHANGE COMPLEX

The soil solution contains a mixture of cations. All of them are subject to attraction to clay surfaces. In the preceding sections, we have seen that the adsorption and exchange of cations depend upon the concentration and the nature of replacing (added) cations. The nature and the composition of the soil solution will change accordingly, depending on the type and amounts of cations being replaced, and on the concentration of the end products as determined by solubility and dissociation. Understanding the changes that can occur in ionic composition is very important in practice. We can then manipulate it through fertilizers and lime addition in favor of plant growth. In order to be able to study this problem, means must be available to formulate the ionic composition in the soil in the presence of charged colloids. A number of equations have been reported for this purpose.

6.6 EMPERICAL EQUATIONS OF CATION EXCHANGE

The Freundlich Equation

The adsorption equation of Freundlich is one method to express ionic composition in the soil solution. It is adaptable to adsorption reactions in a narrow range. The equation is as follows:

$$x = kC^{1/n} \tag{6.1}$$

where

x = amount of cations adsorbed per unit of adsorbent
C = equilibrium concentration of the added cation
k, n = constants

The Langmuir-Vageler Equation

$$\frac{x}{x^o} = \frac{kC}{1 + kC} \tag{6.2}$$

where

x = amount of cations adsorbed per unit weight of exchanger
x^o = total exchange capacity
C = concentration of added cations in moles per liter
k = affinity coefficient

The constant k can be determined as follows (Thomas, 1974), Rearranging Eq. (6.2) gives

$$x\,(1 + kC) = x^o kC$$

$$x + xkC = x^o kC$$

$$x = x^o kC - xkC$$

$$x = kC(x^o - x)$$

$$kC = \frac{x}{x^o - x}$$

therefore

$$k = \frac{x}{C(x^o - x)} \tag{6.3}$$

6.7 MASS ACTION LAW EQUATIONS OF CATION EXCHANGE

Kerr's Equation

Assume that we have the following exchange reactions

$$2Na^+ + Ca\ soil \rightleftharpoons (Na)_2\ soil + Ca^{2+}$$

According to Kerr, the equilibrium is expressed as follows:

$$\frac{[Na^+]^2(Ca^{2+})}{[Ca^{2+}](Na^+)^2} = k \tag{6.4}$$

The sign [ ] denotes adsorbed ions and ( ) denotes free ions in solution. Conforming to the mass action law, the equation says that the ratio of the activity product of reaction products and that of the reactants is constant. By taking the square root, Eq. (6.4) changes into

$$\frac{[Na^+](\sqrt{Ca^{2+}})}{[\sqrt{Ca^{2+}}](Na^+)} = k \tag{6.5}$$

This equation is also known as the *Gapon equation.*

Vanselow's Equation

Vanselow considers the activity of adsorbed cations proportional to the mole fraction of total occupying cations present:

$$\text{Proportion of Na adsorbed} = \frac{[Na^+]}{[Na^+ + Ca^{2+}]}$$

$$\text{Proportion of Ca adsorbed} = \frac{[Ca^{2+}]}{[Na^+ + Ca^{2+}]}$$

By substituting the above in Eq. (6.4) gives

$$\frac{([Na^+]/[Na^+ + Ca^{2+}])^2 \, (Ca^{2+})}{([Ca^{2+}]/[Na + Ca^{2+}]) \, (Na^+)^2} = k$$

Rearranging the above gives Vanselow's equation:

$$\frac{[Na^+]^2(Ca^{2+})}{[Ca^{2+}] \, [Na^+ + Ca^{2+}] \, (Na^+)^2} = k \qquad (6.6)$$

Again [ ] denotes adsorbed ions, and ( ) denotes ions in solution. Krishnamoorty and Overstreet (1949) have developed a similar equation.

6.8 KINETIC EQUATIONS OF CATION EXCHANGE

This type of formulation has been developed by Gapon (1933), Jenny (1936) and by Davis (1945). They are in essence the same formula as expressed by Eq. (6.5).

6.9 EQUATIONS BASED ON THE DONNAN THEORY

A Donnan system is a system composed of solution i and o, separated by a semipermeable membrane (i = inside solution, o = outside solution):

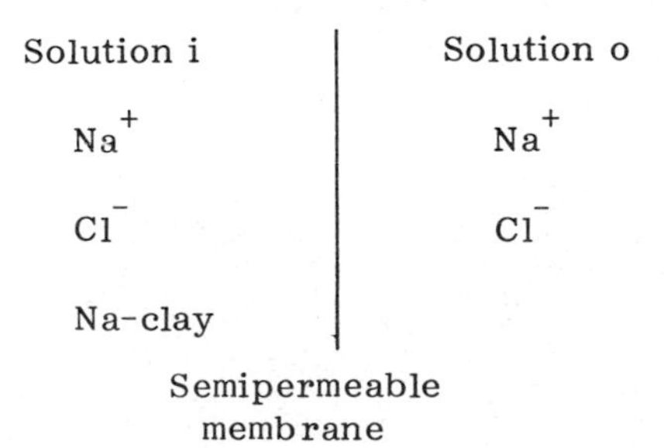

Solution i contains Na^+ and Cl^- and Na-clay, whereas solution o contains only Na^+ and Cl^- ions of different concentrations than those in solution i. The membrane is only permeable to Na^+ and Cl^- ions; therefore only these ions will move and distribute themselves in solutions i and o until equilibrium is reached. At equilibrium the following relation holds:

$$(Na^+)_i\ (Cl^-)_i = (Na^+)_s\ (Cl^-)_s$$

or

$$\frac{(Na^+)_i}{(Na^+)_s} = \frac{(Cl^-)_s}{(Cl^-)_i}$$

Donnan systems are present in soils and are of special importance in soil solution—plant root relationships. They have been extended to cation exchange phenomena and predict essentially the same as the mass action law:

$$\frac{[Na]^2\ (Ca^{2+})}{(Na^+)^2 [Ca^{2+}]} = k$$

Donnan assumes k = 1, therefore the above equation changes into

$$\frac{[Na^+]^2\ (Ca^{2+})}{(Na^+)^2\ [Ca^{2+}]} = 1$$

or (6.7)

$$\frac{[Na^+]}{(Na^+)} = \frac{[\sqrt{Ca}^{2+}]}{(\sqrt{Ca}^{2+})}$$

6.10 EQUATION OF ERIKSSON

Eriksson has combined the Donnan and Vanselow theories and found:

$$\frac{[Na^+]^2\ (Ca^{2+})\ (C)}{(Na^+)^2\ [Ca^{2+}]\ [Na^+ + Ca^{2+}]} = k \qquad (6.8)$$

C = the exchange capacity of the colloid

6.11 EQUATIONS ACCORDING TO THE DIFFUSE DOUBLE-LAYER THEORY

Although considered by many to be the most realistic approach to describe ion exchange equilibria, the double-layer formula developed by Eriksson (1952) is far from simple. It has limited applications and works well for cation exchange equilibrium between Na and Ca but not between other ions. Therefore, for those who are in need of the equation, reference is made to Eriksson (1952) and Lagerwerff and Bolt (1959).

$$\frac{[Na^+]}{[Ca^{2+}]} = \frac{[(1/\beta)/\text{arc sinh}]\ (\beta)^{\frac{1}{2}}\ \Gamma/[(Na^+)/(Ca^{2+})^{\frac{1}{2}} + 4VC\ (Ca^{2+})^{\frac{1}{2}}]}{\Gamma - [(Na^+)/(Ca^{2+})^{\frac{1}{2}}/(\beta)^{\frac{1}{2}}]\ \text{arc sinh}\ (\beta)^{\frac{1}{2}}\ \Gamma/[(Na^+)/(Ca^{2+})^{\frac{1}{2}}] + [4\ VC\ (Ca^{2+})^{\frac{1}{2}}]} R$$

$$R = \frac{(Na^+)}{(Ca^{2+})^{\frac{1}{2}}}$$

$$\beta = 8/100\ RT$$

$$VC = 1$$

$$\Gamma = \text{charge density (mEq/cm}^2\text{) and arc sinh} = \sinh^{-1}$$

The diffuse double-layer equation can be reduced to a Gapon equation.

6.12 SCHOFIELD'S RATIO LAW

From the preceding discussion, it is apparent that although many equations are available to describe cation exchange phenomena, almost all of them have one thing in common. They all stated that the ratios of the products of adsorbed cations and cations free in solution are constant. This can be illustrated by the mass action and the Gapon equations. Both use the formula

$$\frac{[Na^+]\ (\sqrt{Ca^{2+}})}{[\sqrt{Ca^{2+}}]\ (Na^+)} = k$$

in which [ ] denotes adsorbed ions, and ( ) denotes ions in solution. Rearranging the equation gives

$$[Na^+](\sqrt{Ca^{2+}}) = k\ [\sqrt{Ca^{2+}}]\ (Na^+)$$

or (6.9)

$$\frac{(Na^+)}{(\sqrt{Ca^{2+}})} = \frac{1}{k} \frac{[Na^+]}{[\sqrt{Ca^{2+}}]}$$

The latter means that in equilibrium condition, the ratio of cations in solution depends upon the ratio of cations adsorbed on the colloid surface. If the amount of cations adsorbed does not change significantly, or remains constant, the ratio of cations in solution $(Na^+)/(\sqrt{Ca^{2+}})$ is also constant. This is called the *ratio law* by Schofield (1947). The ratio law is used to predict the soil solution concentration as affected by fertilizers and lime application. Ca^{2+} varies by square root, so that a doubling of both Na^+ and Ca^{2+} in the soil will not maintain the same ratio of ions adsorbed.

6.13 FIXATION OF CATIONS

Under certain conditions the adsorbed cations are held so strongly by clays, that they cannot be recovered by exchange reactions. These cations are called *fixed cations*. Although fixation can occur with almost any cation, the most important fixation reaction is with K^+ and NH_4^+ ions. Fixation of K^+ and NH_4^+ occurs by a similar mechanism. Among the several reasons reported for fixation, the most important is the entrapment of the ions in the intermicellar regions of the clays. Expanding lattice clays have octahedral holes of 1.40 Å in their intermicellar surfaces. When K^+ or NH_4^+ penetrates the intermicellar space, they will fit snuggly into the holes. Upon closure of the space, the K^+ or NH_4^+ ions are trapped between the clay layers. They become, then, relatively nonexchangeable and are called *fixed* (Bolt et al., 1976; Rich, 1968; Van der Marel, 1959).

Many soil minerals have been reported to contribute to K^+ and NH_4^+ fixation, e.g., micas, illites, montmorillonites, and vermiculites. Van der Marel (1959) stated that permutites, zeolites, feldspars, and glauconite had the capacity to fix K. Rich (1968) was of the opinion that especially those minerals with strong interlayer charges and wedge zones exhibited sites for high potassium selectivity, in other words, exhibited high K fixation.

Although the quantity of K retained by fixation can assume high proportions, K fixation is currently considered more a disturbing than a harmful reaction. Depending upon the conditions, significant amounts of the fixed K can be released and made available to plant growth. The presence of humic and fulvic acids in soils may accelerate this release (Tan, 1978b). Tisdale and Nelson (1975) are of the opinion that K fixation is a process of conservation in nature. Fixation is of special importance in sandy soils, where K is more likely to be lost rapidly by leaching. Continued application of K or NH_4 fertilizers will decrease K fixation. The addition of K will fill the vacant positions in the clay lattice satisfying in this way the fixation capacity of soils.

6.14 BASE SATURATION

The base saturation is a property closely related to CEC. It is defined as

$$\% \text{ base saturation} = \frac{\Sigma \text{ exchangeable bases (in mEq per 100 g)}}{\text{CEC}} \times 100\%$$

Example: Assume that in the analysis of soils the following concentration of bases and CEC value were obtained:

Exchangeable bases, mEq per 100 g soil	
Ca	10
Mg	5
K	10
Na	5
Σ exchangeable bases	= 30

The CEC of the soil, determined separately, is 50 mEq per 100 g. Therefore:

$$\% \text{ base saturation} = \frac{30}{50} \times 100 = 60\%$$

A positive correlation exists between percent base saturation and soil pH. Generally, we can see that the base saturation is high, if soil pH is high. Consequently, arid region soils are usually higher in base saturation than soils in humid regions. Low base saturation means the presence of a lot of H^+ ions.

The base saturation is frequently considered to be an indication of soil fertility. The ease with which adsorbed cations are released to plants depends on the degree of base saturation. A soil is considered very fertile if the percent base saturation is $\geqslant 80\%$, medium fertile if percent base saturation is 80 to 50%, and nonfertile if percent base saturation is $\leqslant 50\%$. A soil with a percent base saturation of 80% will release the exchangeable bases more easily than the same soil with a percent base saturation of 50%. Liming is the common means by which the percent base saturation of soils is increased.

7

ANION EXCHANGE

7.1 POSITIVE CHARGES

In a preceding section, it was indicated that under certain conditions the soil colloids may also carry positive charges. This is especially true for the Fe and Al oxide minerals and amorphous soil colloids. Positive charges can also occur at the edges of clay minerals. This kind of charge is usually significant at pH values below the isoelectric point or zero point of charge. The broken edge surface of an octahedral sheet has, therefore, a positively charged double layer at low pH. The double layer becomes increasingly more positive with decreasing pH.

7.2 ADSORPTION OF ANIONS BY SOIL COLLOIDS

Two types of adsorption of anions by soil colloids are recognized, i.e., negative and positive adsorption.

Negative Adsorption

Negative adsorption of anions occurs at a colloidal surface possessing a negative charge. Because of the latter, cations are attracted and concentrated at the colloid surface. On the other hand, anions are expelled from the double layer formed on the negative charged surface. This exclusion of anions is called *negative adsorption*. Therefore, the bulk solution contains more anions than the solution in the interface. The amount excluded is reported to be a small part of the CEC. Bolt (1976) shows that under conditions prevailing in soil, the negative adsorption of anions is approximately 1 to 5% of the CEC. In contrast, negative adsorption can amount to 15% in saline soils.

Positive Adsorption

Positive adsorption of anions is the adsorption and concentration of anions on the positively charged surfaces or edges of soil colloids. In this case, negative adsorption of cations, i.e., repulsion of cations by the positive charge, occurs.

The anion exchange capacity (AEC) of soils is usually smaller than the CEC. It is dependent on changes in electrolyte levels and on soil pH. It is also limited to special types of clays.

As is the case with cations, lyotropic series of anions are also available. Bolt (1976) reported a decreasing order of preferential adsorption among the following anions:

$$SiO_4^{4-} > PO_4^{3-} >> SO_4^{2-} > NO_3^{-} \approx Cl^{-}$$

The lyotropic series above indicates that SiO_4^{4-} and PO_4^{3-} ions are strongly adsorbed. SO_4^{2-} and NO_3^- ions are adsorbed in considerably lower concentrations or are often not adsorbed. At a neutral soil reaction, or pH > 6, positive adsorption of SO_4^{2-} and Cl^- is very small. These two anions are generally subject to negative adsorption, in contrast to phosphate ions which are adsorbed more by positively charged surfaces or edges of clay minerals:

$$\underset{\text{(Clay)}}{Al - OH} + H_2PO_4^- \rightleftharpoons Al - H_2PO_4 + OH^-$$

The reaction above is prevalent in acid soils. It results in a strong bond between the phosphate ion and the octahedral Al. Frequently, only part of the phosphate can be recovered by desorption analysis.

7.3 PHOSPHATE FIXATION AND RETENTION

Phosphate anions can be attracted to soil constituents with such a bond that they become insoluble and difficultly available to plants. The process above is called *phosphate fixation* or *phosphate retention*. Many authors use the terms *fixation* and *retention* interchangeably. However, Tisdale and Nelson (1975) are of the opinion that *retention* refers to that part of adsorbed phosphorus that can be extracted with dilute acids. This fraction is relatively available to plants. The term *fixed*, on the other hand, is reserved for the portion of soil phosphorus, which is not extractable by dilute acids. This portion of phosphorus is not readily available to plants. Under certain conditions, the distinction between phosphorus fixation and retention is rather obscure.

Phosphate Retention

Acid soils usually contain significant amounts of soluble and exchangeable Al^{3+}, Fe^{3+}, and Mn^{2+} ions. Phosphate when present may be adsorbed to the colloid surface with these Al^{3+}, Fe^{3+}, and Mn^{2+} ions serving as a bridge. This phenomenon is sometimes called *coadsorption*. The phosphate retained in this way is still available to plants. Such a reaction can also take place with Ca-saturated clays. Evidence has been shown that Ca clay adsorbs large amounts of phosphate. The Ca^{2+} ions form the linkage between the clay and phosphate ions:

$$\boxed{\text{Clay}} - Ca - H_2PO_4$$

According to the mass action or Gapon equation, the ionic concentration at the surface of the clay is dependent on that of the bulk solution:

$$[\sqrt{Ca^{2+}}][H_2PO_4^-] = (\sqrt{Ca^{2+}})(H_2PO_4^-)$$

in which [ ] denotes adsorbed species, and ( ) again equals free species. The phosphate ions can also enter into a chemical reaction with the free metal ions above:

$$Al^{3+} + 3H_2PO_4^- \rightarrow Al(H_2PO_4)_3\downarrow$$

The product formed is difficultly soluble in water and precipitates from solution. With the passage of time the Al phosphate precipitates become less soluble and less available to the plant.

The lower the soil pH, the greater the concentration of soluble Al, Fe, and Mn; consequently, the larger the amount of phosphorus retained in this way.

Phosphate Fixation

As stated above, fixation renders phosphate insoluble in water and relatively nonavailable to plants. The fixation reaction can occur between phosphate and Al or Fe hydrous oxides or between phosphate and silicate minerals.

Many soils contain high amounts of Fe and Al hydrous oxide clays, especially the highly weathered Ultisols in the United States and the Oxisols in the tropics. These clays react rapidly with phosphate forming a series of difficultly soluble hydroxy phosphates:

$$\begin{array}{c} H \\ O \\ | \\ Al-OH \\ | \\ O \\ H \end{array} + H_2PO_4^- \rightarrow \begin{array}{c} H \\ O \\ | \\ Al-H_2PO_4 \\ | \\ O \\ H \end{array}$$

Insoluble

The amount of phosphorus fixed by this reaction usually exceeds that fixed by phosphate retention. Such a reaction is not limited to Al and Fe hydrous oxide clays, but Mn hydrous oxide clays and amorphous clays are also known to have considerable phosphate fixing capacities. In contrast to phosphate retention that occurs mainly in acid soil condition, phosphate fixation by hydrous oxide clays occurs over a relatively wider pH range.

The products formed by both retention and fixation reactions are frequently not pure Al or pure Fe phosphates. The ultimate end product of the reaction between aluminum hydroxides and phosphates is called *variscite* ($AlPO_4 \cdot 2H_2O$) and that of Fe and phosphate is known as *strengite* ($FePO_4 \cdot 2H_2O$). A series of intergrades between variscite and strengite is usually present in soils and is called the variscite-strengite isomorphous series (Lindsay et al., 1959).

Another type of phosphate fixation is the reaction between phosphate and silicate clays. Especially soil clays exhibiting exposed OH groups, such as the kaolinitic groups, have a strong affinity to phosphate ions. Generally, clays with low $SiO_2:R_2O_3$ (sesquioxide) ratios have a higher phosphate fixing capacity than clays with high $SiO_2:R_2O_3$ (sequioxide) ratios.

The phosphate ions react rapidly with octahedral Al by replacing the OH groups located on the surface plane of the mineral. This type of reaction is also prevalent in acid conditions.

Phosphate Fixation in Alkaline Soils

Fixation of phosphate is not limited to acid conditions only, but also occurs readily in alkaline soil reactions. Many alkaline soils contain high amounts of soluble and exchangeable Ca^{2+} and frequently $CaCO_3$. Phosphate is reported to react with both the ionic and the carbonate form of Ca. The latter can be illustrated as follows:

$$3\,Ca^{2+} + 2PO_4^{3-} \rightarrow Ca_3(PO_4)_2\downarrow$$

(insoluble)

$$3\ CaCO_3 + 2PO_4^{3-} \rightarrow Ca_3(PO_4)_2 + 3\ CO_2\uparrow$$

(insoluble)

Other forms of insoluble Ca phosphate can also be formed by this type of reaction between calcium and phosphate, e.g., hydroxy-, oxy- and Ca-fluoroapatite [$Ca_5(PO_4)_3F$].

Such type of fixation is a serious problem in arid region soils of the western part of the United States. However, it can also become significant in the humid region of the eastern seaboard of the United States when the soils receive high applications of lime. In these soils or under such conditions application of P fertilizers generally gives low plant growth response. Phosphate fixation cannot be avoided entirely, but it may be reduced by addition of competing ions for fixing sites. Organic anions from stable manure and silicates are reported to be very useful in reducing P fixation (Bolt, 1976).

7.4 SOIL REACTION AND AVAILABILITY OF INORGANIC PHOSPHATES

As discussed above, insoluble Al and Fe hydroxy phosphates are stable in acid conditions, whereas Ca phosphates are prevalent in alkaline conditions. At pH 3 to 4, solubility of the Al and Fe hydroxy phosphates is considered very low. However, with increasing pH levels, solubility of these phosphate compounds increases and reaches a maximum at approximately pH 6.5. Above this pH level, the Al- and Fe-hydroxy compounds decrease again in solubility (see Figure 7.1). On the extreme alkaline range, pH 8.0, Ca phosphate is in an insoluble form. By decreasing the pH this compound also becomes slightly soluble, and maximum solubility is reached at pH 6.5.

Therefore, it appears that at pH 6.5, the soil may contain the maximum amounts of phosphate that can be solubilized from all the insoluble inorganic phosphate forms present in soils. The forms of phosphate ions present also depend on soil pH. It is known that in acid conditions the $H_2PO_4^-$ ions prevail, in alkaline conditions the HPO_4^{2-} ions are dominant, while at pH 6.5, $H_2PO_4^-$, HPO_4^{2-}, and PO_4^{3-} can exist in combination in the soil solution.

7.5 PREDICTION OF PHOSPHATE ION CONCENTRATION ACCORDING TO DOUBLE-LAYER THEORIES

From the above, it is apparent that the concentration of phosphate ions in the soil solution depends upon a number of factors. It is usually affected by soil pH, by some member of the variscite-strengite isomorphous series, and by a member of the Ca-hydroxy

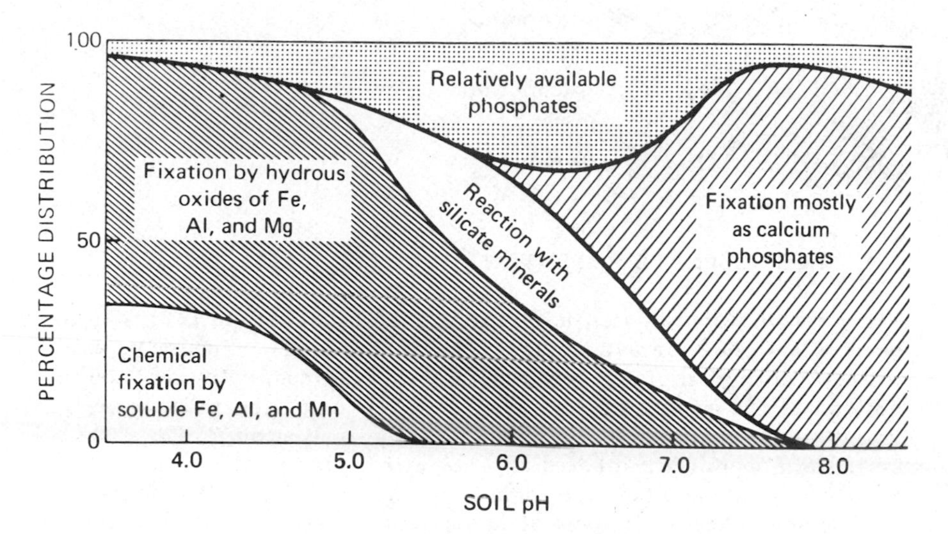

Figure 7.1 Phosphate availability and fixation as related to soil pH. (From N.C. Brady, *The Nature and properties of soils.* 8th ed., Macmillan, New York, Copyright 1974 by Macmillan Publ. Co., Inc.)

phosphate series. According to Schofield's ratio law, the following relations are valid:

$$(H^+) = \frac{K_w}{(OH^-)} \tag{7.1}$$

$$\frac{[\sqrt{Ca^{2+}}]}{[H^+]} = \frac{(\sqrt{Ca^{2+}})}{(H^+)} = k_1 \tag{7.2}$$

$$\frac{(Al^{3+})^{1/3}}{(Ca^{2+})^{1/2}} = k_2 \tag{7.3}$$

Lindsay et al. (1959) assume that if variscite, $Al(OH)_2H_2PO_4$, will dissociate completely into its component ions, we can write the solubility product (K_{sp}) of variscite as follows:

$$K_{sp} = (Al^{3+})(OH^-)^2(H_2PO_4^-)$$

Using Eq. (7.1) to (7.3) for substitution into the equation for K_{sp}, Lindsay et al. (1959) stated,

$$(\sqrt{Ca^{2+}})\,(H_2PO_4^-) = \text{constant}$$

$$(H^+)\,(H_2PO_4^-) = \text{constant}$$

$$(Al^{3+})^{1/3}\,(H_2PO_4^-) = \text{constant}$$

7.6 THE PHOSPHATE POTENTIAL

The term *phosphate potential* is introduced by Schofield (1955) to be used as an index for availability of soil phosphorus. Similarly, as with the definition of soil water potential, the phosphate potential is defined as the amount of work that must be conducted to move reversibly and isothermally an infinitesimally small amount of a phosphate ion from a pool of phosphates at a specified location at atmospheric pressure to the point under consideration. Schofield (1955) was of the opinion that a labile pool of phosphates existed in the soil from which plant roots could draw their P needs. By measuring the pH and the total P concentration in 0.01 M $CaCl_2$ solution, the solubility of phosphate can be expressed as

$$(Ca^{2+})^{1/2}\,(H_2PO_4^-)$$

By taking -log, this product changes to

$$-\tfrac{1}{2}\log(Ca^{2+}) + [-\log(H_2PO_4^-)]$$

or

$$\tfrac{1}{2}\,pCa + pH_2PO_4 \qquad (7.4)$$

in which p = -log. Formula (7.4) is called the *phosphate potential.* In analogy to the water potential, a low phosphate potential suggests high availability, whereas a high phosphate potential refers to lower availability of P to plants. Due to a number of technical difficulties, the use of phosphate potentials as an index of availability of P has not been tested. However, since expression (7.4) is a solubility product, it is useful in estimating solubility of phosphorus in soil solutions. Since availability of P to plants is related to solubility of P, the phosphate potential can be used to make indirect predictions as to the phosphate availability to plants.

8

SOIL REACTION

8.1 DEFINITION AND IMPORTANCE

The *soil reaction* is a term used to indicate the acid-base reactions in soils (Tisdale and Nelson, 1975; Brady, 1974). A number of soil processes are affected by the soil reaction. Many soil chemical and biochemical reactions can occur only at specific soil reactions. The rate of decomposition of soil minerals and organic matter is influenced by the soil reaction. Formation of clay minerals depends on the soil reaction. Plant growth is also affected either directly or indirectly by the acid-base reactions in soils. The latter may influence plant growth indirectly through its effect on solubility and availability of plant nutrients. The changing phosphate concentration with soil pH, as discussed earlier, is a good example. Directly, H^+ions are reported to have a toxic effect on plants when present in high concentration.

The colloidal particles can also behave as an acid or a base. Hydrogen or Al-saturated clays usually behave as an acid and may react with bases.

8.2 ACID-BASE CHEMISTRY

In order to understand and manipulate the soil acid-base system, it is important to first define what an acid or a base is. Three major concepts of acids and bases are available, i.e., the (1) Arrhenius, (2) Brønsted, and (3) Lewis concepts.

Arrhenius Concept

The concept of Arrhenius developed between 1880 and 1890 stated that an acid is a compound that contains hydrogen. In aqueous solution the acid yields hydrogen ions (H^+). A base is a compound that produces hydroxyl ions (OH^-) in solutions. The Arrhenius concept is essentially valid only for the definition of acids. Almost all acids contain hydrogen. However, Arrhenius' definition of a base limits the base only to compounds with hydroxyl ions. It is currently known that ammonia (NH_3) and many other organic substances exhibit characteristics of base compounds.

Brønsted-Lowry Concept

Brønsted and Lowry defined separately, in 1923, that an *acid* is a compound capable of donating a proton (proton donor). On the other hand, any compound capable of accepting a proton is considered a *base* (proton acceptor). If we take as an example HCl, which is usually considered an acid, then according to the Brønsted-Lowry theory, the compound can donate a proton. Therefore, HCl is an acid. However, after dissociation the remaining Cl^- ion is then a base, because it can accept a proton. This type of acid-base pair is called a conjugate pair with the Cl^- being the conjugate base of the acid HCl.

If HCl is dissolved in water, the following reaction occurs:

$$HCl + H_2O \rightarrow H_3O^+ + Cl^-$$

H_3O^+ is called a *hydronium ion*. Since hydronium is formed by adsorption of a proton by the water molecule, water is a proton acceptor. Consequently water can be considered a base. The reaction to the left ($H_2O + Cl^-$) will not occur since H_2O is a stronger proton acceptor than proton donor.

Lewis Concept

According to this theory, also developed in 1923, an acid is a compound that can accept an electron pair. A base is a compound that can donate an electron pair. The following serves as an example:

$$H^+ + :\ddot{\underset{..}{O}}:H^- \rightarrow H:\ddot{\underset{..}{O}}:H$$

In the above, the H^+ ion accepts an electron pair. The proton is then the acid and the bonding is called a covalent bond. The hydroxide ion donates the electron pair and is considered the base. Another example is as follows:

$$H^+ + \begin{matrix} & H & \\ : & \ddot{\underset{..}{N}} & :H \\ & H & \end{matrix} \rightarrow H\begin{matrix} & H & \\ : & \ddot{\underset{..}{N}} & :H \\ & H & \end{matrix}$$

Again, the H^+ ion accepts an electron pair and is the acid. NH_3 is then the base, since it donates the electron pair.

In studying the three theories above, it can perhaps be noted that in the soil solution at one time both the Brønsted-Lowry and the Lewis theories can be applied. However, at another time the Arrhenius and Brønsted theories are more suitable to describe the condition. Reactions with clay minerals perhaps follow the Brønsted-Lowry concept more, while complex reactions involving organic matter apply only to the Lewis theory.

8.3 FORMULATION OF SOIL ACIDITY AND ALKALINITY

Soil pH

The theories of Arrhenius and Brønsted in combination must be applied to characterize acid and alkaline conditions in soils. In acid soils, more H^+ than OH^- ions are present. On the other hand, a basic soil has in its soil solution more OH^- than H^+ ions. To characterize these conditions the term *soil pH* is used. The term pH was introduced by Sörensen in 1909 and is defined as

$$pH = \log \frac{1}{A_{H^+}} = -\log A_{H^+} \tag{8.1}$$

where A_{H^+} = activity of H^+ ions

However, frequently it is more convenient to use H^+ ion concentration rather than activity, and Eq. (8.1) becomes

$$pH = \log \frac{1}{(H^+)} = -\log (H^+)$$

Application of this pH concept in the dissociation of pure water gives the following relationship. As discussed earlier the ion product of water is

$$K_w = C_{H^+} \times C_{OH^-} = 10^{-14} \text{ at } 25°C$$

By taking the -log, the equation above changes into

$$-\log C_{H^+} - \log C_{OH^-} = -\log 10^{-14}$$

or $\quad -\log (H^+) - \log (OH^-) = 14$

or $\quad pH + pOH = 14 \qquad (8.2)$

in which p = -log; therefore pOH = -log of the hydroxyl ion concentration. It (pOH) is calculated in a similar manner as pH.

Equation (8.2) states that pH + pOH = constant and conversion of one into the other is a simple matter. Therefore, in describing soil acidity or alkalinity it is not necessary to determine both pH and pOH. If pH is known, pOH can be calculated using Eq. (8.2).

Acidity Constant

According to the Brønsted-Lowry concept the following acid-base relationship is valid:

$$\text{Acid} \rightleftharpoons H^+ + \text{base}$$

Since this is a dissociation reaction, at equilibrium we therefore have

$$K_A = \frac{(H^+)(\text{base})}{(\text{acid})}$$

in which K_A is called the *acidity constant*. Application of the pH concept in the equation above gives

$$pK_A = \frac{\text{pH } [-\log(\text{base})]}{-\log(\text{acid})}$$

or $$\text{pH} = pK_A + \log \frac{(\text{base})}{(\text{acid})}$$

The latter means that for a given ratio of concentration (or activities) of an acid and its conjugated base, the pH has a fixed value (Novozamsky et al., 1976).

Acid Strength and Ion Pairs

The strength of an acid HB depends on the reaction of the acid with the solvent:

$$\underset{\text{acid}}{HB} + \underset{\text{solvent}}{HS} \rightarrow \underset{\text{ion pair}}{H_2S^+B^-}$$

The reaction above results in ionization of the acid and conjugated base. In most electrolytes, the resulting component ions do not completely dissociate from each other. The cations and anions are strongly attracted to each other and a large part behaves as if it is not ionized. These ions present in association are called *ion pairs* (Davies, 1962). The degree of ionization is dependent on the relative alkaline strengths of the conjugated base and the solvent.

The ion pair will dissociate into its component ions:

$$H_2S^+B^- \rightleftharpoons H_2S^+ + B^-$$

(dissociation)

The degree of dissociation depends on the dielectric constant of the solvent. (The dielectric constant is the ratio of the capacity of an electric condenser in a vacuum and in the solvent, $\varepsilon = C_V/C_S$.)

By combining the two reactions above the dissociation constant (K_A) of HB can be written as

$$K_A = \frac{(H_2S^+)(B^-)}{(HB) + (H_2S^+B^-)}$$

The value for K_A is used as a quantitative measurement of the strength of the acid.

8.4 CONCEPTS OF SOIL ACIDITY

Soil pH Range

Based on the relative degree of acidity, the soils are divided into several acidity or alkalinity classes as shown in Figure 8.1. Acid soils are usually common in humid regions. In these soils, the concentration of H^+ ions exceeds that of OH^- ions. These soils may contain large amounts of soluble Al, Fe, and Mn. Alkaline soils occur mostly in semiarid to arid regions. Here the OH^- ions are dominant over the H^+ ions. Due to the alkaline reaction the soils contain low amounts of soluble Al, Fe, and Mn.

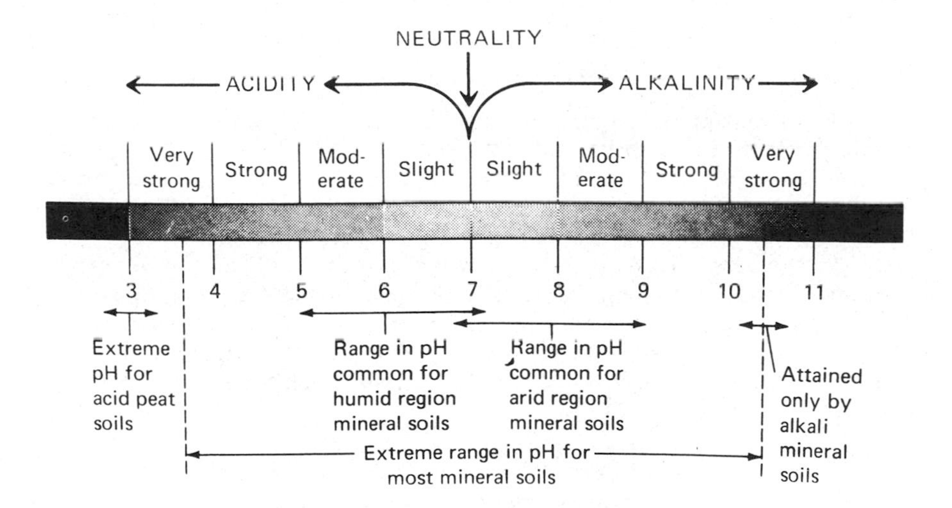

Figure 8.1 Soil pH ranges or soil reaction classes. (From N.C. Brady, *The Nature and Properties of Soils*. 8th edition. Macmillan New York, Copyright 1974 by Macmillan Publ. Co., Inc.)

Active versus Potential Soil Acidity

A number of compounds contribute to the development of acid or basic soil reactions. Inorganic and organic acids, produced by the decomposition of soil organic matter, are common soil constituents that may affect soil acidity. Respiration of plant roots yield CO_2 that will produce H_2CO_3 in water. Water is another source for a small amount of H^+ ions. A large portion of the H^+ ions present in soils will be absorbed by the clay complex as exchangeable H^+ ions:

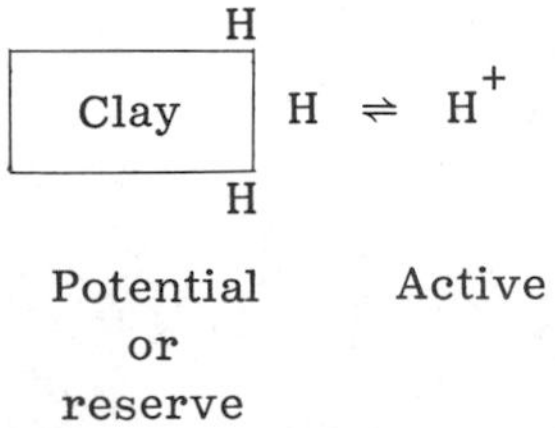

$$\boxed{\text{Clay}}\begin{matrix} H \\ H \\ H \end{matrix} \rightleftharpoons H^+$$

Potential or reserve   Active

These exchangeable H^+ ions dissociate into free H^+ ions. The degree of ionization and dissociation into the soil solution determines the nature of soil acidity. The exchangeable H^+ ions are the reason for the development of potential or reserve soil acidity. The magnitude of the latter can be determined by titration of the soil. The free H^+ ions create the active acidity. Active acidity is measured and expressed as soil pH. This is the type of acidity upon which plant growth reacts.

Nonselective and Preferential Adsorption of H^+ Ions by Soils

Bolt (1976) distinguishes two types of adsorption of H^+ ions by soil colloids, i.e., nonselective and preferential adsorption. Soil colloids, such as clay, adsorb H^+ ions by the nonselective process. They accumulate on the charged surface as a swarm of counter ions. The relative proportion can be estimated with the Gapon equation.

On the other hand, organic colloids exhibit preferential adsorption of H^+ ions. The organic compounds contain acidic groups that are highly selective for association with protons. The adsorbed H^+ ions are, then, considered part of the group, or the surface. Bolt (1976) considered them more difficultly exchangeable against other cations.

8.5 THE ROLE OF Al IN SOIL ACIDITY

Most clay particles interact with H^+ ions. Evidence is available that a hydrogen saturated clay undergoes a spontaneous decomposition. The hydrogen ion penetrates the octahedral layer and replaces the Al atoms.

The Al released is then adsorbed by the clay complex and a H-Al-clay complex is formed rapidly. The Al^{3+} ions may hydrolyze and produce H^+ ions:

$$\boxed{\text{clay}}\,\text{Al} + 3H_2O \rightarrow Al(OH)_{3\downarrow} + H\,\overset{H}{\underset{H}{\boxed{\text{clay}}}} \rightleftharpoons H^+$$

The above reaction contributes towards increasing the H^+ ion concentration in soils.

8.6 BUFFERING CAPACITY OF SOILS

Chemically a buffer solution is defined as one that resists a change in pH on addition of acid or alkali. Buffer solutions contain compounds that react with both acid or base so that the H^+ ion concentration in the solution remains constant.

In soils the clay and humic fractions act as a buffer system. As discussed previously, the soil cation exchange complex creates the development of potential and active acidity. The potential acidity will maintain the equilibrium with the active acidity. If the active H^+ ion concentration is neutralized by the addition of lime, the potential acidity will release exchangeable H^+ ions into the soil solution to restore the equilibrium, and no change in soil reaction occurs until the reserve in H^+ is exhausted. The magnitude of the potential acidity usually far exceeds that of the active acidity. Brady (1974) reported that in sandy soils, reserve acidity (H^+ on the exchange complex) was 1000 times greater than active acidity. In clay soils high in organic matter, reserve acidity was even 50,000 to 100,000 times greater than active acidity. Therefore, buffering capacity is greater in clay soils than in sandy soils. The larger the buffer capacity, the larger the amounts of lime needed to raise the soil pH to the desired level.

The concept of buffering capacity of soils is not limited to the soil's resistance to changes in soil reaction. The soil can also act as a filter for dissolved and colloidal contaminants. It may act as a sieve, or during the passage through the top soil, the aerated condition may oxidize and mineralize, in particular, the organic compounds. The ions released by mineralization are adsorbed by the soil adsorption complex and prevented from reaching the ground water.

8.7 ELECTROMETRIC MEASUREMENT OF SOIL pH

The determination of soil pH is conducted today with the glass electrode. It consists of a thin glass bulb containing dilute HCl, into which is

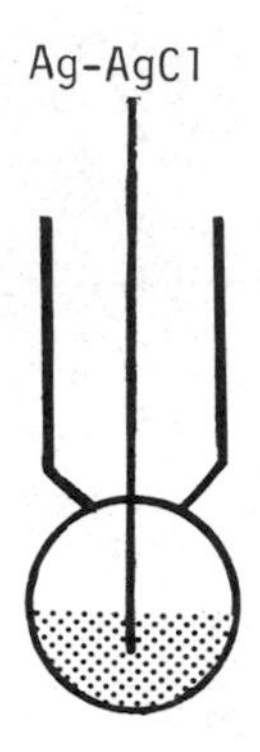

inserted an Ag-AgCl wire, serving as the electrode with a fixed voltage. When the glass bulb is immersed in a solution, a potential difference develops between the solution in the bulb and the soil solution outside the bulb. This potential E is formulated by the *Nernst equation:*

$$E = \frac{RT}{nF} \log \frac{K}{M^{n+}}$$

where

R = gas constant
T = absolute temperature
n = valence
F = Faraday constant
K = constant
M = activity of ions to be measured

E is called the *half-cell potential* and cannot be measured alone. If the glass electrode is placed against a reference electrode (usually the calomel electrode), the potential difference between the two ($E - E_{cal}$) is measurable.

Before any pH measurement, the two electrodes have to be placed first in a solution of known pH (e.g., H^+ ion concentration = 1 g/l). This is called *standardizing the electrodes* and the pH meter. The overall potential of the total cell E_o equals $E - E_{cal}$, and is as follows:

$$E_o = \frac{RT}{F} \log \frac{K}{1} - E_{cal} \tag{8.4}$$

If the two electrodes are now placed in the solution with the unknown H^+ ion concentration, the potential E_c is

$$E_c = \frac{RT}{F} \log \frac{K}{(H^+)} - E_{cal} \tag{8.5}$$

Substracting Eq. (8.4) from Eq. (8.5) gives

$$E_c - E_o = \frac{RT}{F} \left(\log \frac{K}{(H^+)} - \log K\right)$$

$$= \frac{RT}{F} \log \frac{1}{H^+}$$

or $$E_c - E_o = \frac{RT}{F} \text{ pH}$$

For H^+ ions, $RT/F = 0.0591$ at 25°C; therefore

$$\text{pH} = \frac{E_c - E_o}{0.0591} \tag{8.6}$$

8.8 SUSPENSION EFFECT IN SOIL pH MEASUREMENT

In the pH measurement the reference and indicator electrodes are immersed in a heterogeneous soil suspension composed of dispersed solid particles in an aqueous solution. If the solid particles are allowed to settle down, the pH can be measured in the supernatant liquid or in the sediment. Placement of the electrode pair in the supernatant usually gives a higher pH reading than placement of the electrodes in the sediment. This difference in soil pH reading is called the *suspension effect*.

Stirring the soil suspension before measurement will not solve the problem, since the latter procedure gives unstable readings.

According to Bolt et al. (1976), the reference electrode should be placed in the supernatant. The position of the glass electrode is considered immaterial at equilibrium condition.

8.9 LIME POTENTIAL

It is apparent from the discussions above that measurements of soil pH can give highly variable results since they are affected by a number of factors, e.g., suspension effect, soil-water ratios, and electrolyte levels. Schofield and Taylor (1955) proposed the use of 0.1 M $CaCl_2$ solutions for obtaining stable readings in pH measurements.
However, instead of using single ion activity measurements, Schofield and Taylor (1955) suggested the use of ion activity ratios for determination of soil acidity and base saturation. If the soil exchange complex

is saturated with both H^+ and Ca^{2+} ions, at equilibrium Schofield's ratio law says that

$$\frac{(H^+)}{(\sqrt{Ca^{2+}})} = \text{constant}$$

By taking -log, the ratio above changes to

$$-\log \frac{(H^+)}{(\sqrt{Ca^{2+}})} = \text{constant}$$

$$-\log (H^+) - [-\log (\sqrt{Ca^{2+}})] = \text{constant}$$

or $$\text{pH} - \frac{1}{2}\,\text{pCa} = \text{constant} \tag{8.7}$$

This equation is called the *lime potential.* It characterizes the composition of the exchange complex with respect to its saturation by H^+ and Ca^{2+} ions. Many of the methods for determination of lime requirement take into consideration the lime potential.

8.10 THE NEED FOR ACID SOIL REACTIONS

In most cases soil fertility is improved by liming acid soils to pH 6 to 7. Most plants grow well in this pH range. At this soil reaction, available Ca, Mg, and P concentrations are adequate for plant growth. The levels of micronutrient contents in the soil solution are sufficient. Both fungi and bacterial activity are also present.

However, in certain cases it is desirable to maintain a strong to moderately acid condition for plant growth. Some ornamental plants, e.g., azaleas and rhododendrons, required acid soil reactions for optimum growth. Pine trees also grow better in acid soils. Other crops that are grown in acid soils are blueberries, pineapples, and Irish potatoes. With potatoes, the acid condition will reduce the development of potato scab, a disease caused by actinomycetes.

A number of compounds can be used to maintain or intensify soil acidity. Acid organic matter, e.g., pine needles, or chemical compounds such as S powder or $FeSO_4$ can be mixed with the soil to produce the acid reaction.

8.11 SOIL REACTIONS IN SALINE AND SODIC SOILS

Salinization

Saline and sodic soils, today called *Aridisols*, are soils of arid regions where the average precipitation is less than 500 mm (20 in.) annually.

The amount of H_2O coming from the precipitation is insufficient to neutralize the amount of H_2O lost by evaporation and evapotranspiration. As the water is evaporated in the atmosphere, the salts are left behind in the soil. The process of accumulation of soluble salts in these soils is called salinization. The salts are mostly NaCl, Na_2SO_4, $CaCO_3$, and/or $MgCO_3$. In the past, the soils developed were called *saline soils, white alkali soils,* or *solonchaks.* They belong to the zonal type of soils. Salinization can also occur locally and develops the intrazonal type of saline soils, e.g., soils reclaimed from the sea bottom and soils in coastal areas affected by the tide.

Sodication and Alkalinization

The addition of salts to the soil may result in saturating the soils exchange complex with Na. The process of progressively increasing the Na saturation of the soils exchange complex is called *sodication.* The soils formed are called *sodic soils, solods, solonetz,* or *black alkali soils.* If these soils occur only in small areas (in small localized spots), they are often called *slick spots.*

Kamphorst and Bolt (1976) reported that sodication does not necessarily yield a rise in soil pH. Many sodic soils are neutral in reaction, whereas a number of solonetzic soils are even acid in reaction. In sodic soils with a neutral soil reaction, the Na salts are neutral salts such as NaCl.

The strong alkaline reaction (pH = 10) of most sodic soils is caused by alkalinization. The latter is due to hydrolysis of Na^+ ions or Na_2CO_3 compounds:

$$Na_2CO_3 + 2H_2O \rightleftharpoons 2Na^+ + 2OH^- + H_2CO_3$$

The OH^- ions produced will increase the soil pH, while the Na^+ is saturating the exchange complex. The latter may in turn undergo hydrolysis which also contributes toward increasing the OH^- ion concentration in the soil:

$$\boxed{\text{Clay}}\,Na + H_2O \rightleftharpoons \boxed{\text{Clay}}\,H + Na^+ + OH^-$$

8.12 CHEMICAL CHARACTERIZATION OF SALINE AND SODIC SOILS

It is apparent from the above that soil pH is not a good method for characterization of these soils. The saline soils have a soil pH = 8.5 or lower. The sodic soils may possess a soil pH = 10, but some of the soils may be neutral; whereas others are acid in reaction. To distinguish saline and sodic soils from other soils, the U.S. Salinity laboratory (Richards, 1954) proposed to use as criteria soluble salt and exchangeable Na^+ content. These parameters are expressed in terms of (1) electrical conductivity (EC_e) for salt content and (2) exchangeable sodium percentage (ESP) for exchangeable Na^+ content. The salinity of the soil is measured by measuring the EC_e in mmho/cm of the soils saturated extract. The latter is obtained by suction and filtration of a water-saturated soil paste. The exchangeable sodium percentage is calculated using the formula as follows:

$$ESP = \frac{\text{exchangeable Na}^+ \text{ ions}}{\Sigma \text{ exchangeable cations}} \times 100\%$$

Based on ESP and EC_e values three groups of soils are recognized: (1) saline soils, (2) saline-alkali soils, and (3) nonsaline-alkali (sodic) soils. The saline soils are characterized by $EC_e > 4$ mmho/cm at 25°C, and ESP < 15%. Dispersion of saline soils occurs at ESP = 15%. The soil pH is ordinarily less than 8.5. Because of the presence of excess salts and low amounts of Na^+ in exchange position, these soils are usually in a flocculated state and their permeability is considered to be equal or higher than the two other soils. The saline-alkali soils are soils with $EC_e > 4$ mmho/cm at 25°C, and ESP > 15%. These soils have both free salts and exchangeable Na^+. As long as excess salts are present, the soil is flocculated and the pH is normally $\leqslant 8.5$. When the soils are leached, the free salt content decreases, and the soil reaction may become strongly alkaline (pH > 8.5) because of hydrolysis of the exchangeable Na^+. Nonsaline-alkali soils are characterized by $EC_e < 4$ mmho/cm at 25°C, and ESP > 15%. Most of the Na^+ is in exchangeable form, and very small amounts of free salts are present in the soil solution. The soil pH ranges from 8.5 to 10.0. As a result of irrigation, strong alkaline conditions may develop in these soils and pH values reaching 10 are common.

The selection of the critical value for EC_e at 4 mmho/cm was reported to be based on the expected salt damage to crops. The EC_e value of 4 mmho/cm originated with Scofield in 1942, who considered the soil to be saline at 4 mmho/cm or above. At the latter values the yield of many crops is restricted. Kamphorst and Bolt (1976) indicated that an EC_e of 4 mmho/cm corresponded to an osmotic pressure at field capacity of 5 bars. At EC_e values between 2 to 4 mmho/cm, only very sensitive crops will be affected, while at values below 2 mmho/cm the effect of salinity is negligibly small (see Figure 8.2). The decision to use an ESP value of 15% is very arbitrary, since no sharp changes in soil properties have been observed as the degree of saturation of the exchange complex with Na^+ ions is increased. Moreover, different crops will react differently to the same ESP value (Kamphorst and Bolt, 1976; Richards, 1954). The U.S. Salinity Laboratory has used by history and experience the ESP = 15% as a boundary limit to distinguish nonalkali from alkali soils.

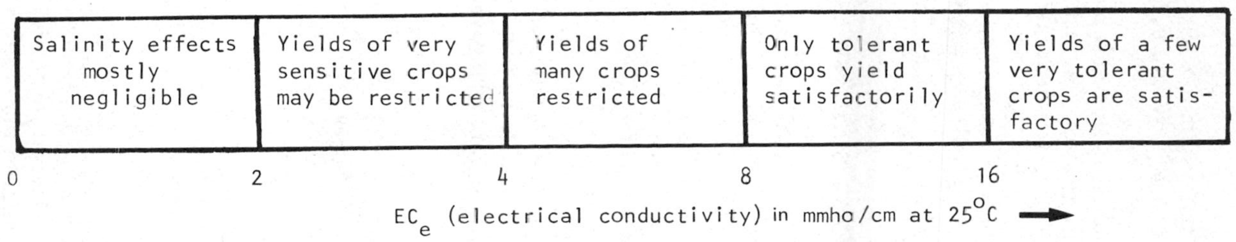

Salinity effects mostly negligible	Yields of very sensitive crops may be restricted	Yields of many crops restricted	Only tolerant crops yield satisfactorily	Yields of a few very tolerant crops are satisfactory

0 2 4 8 16

EC_e (electrical conductivity) in mmho/cm at 25°C →

Figure 8.2 The effect of degree of salinity, as expressed in EC_e values, on yields of crops, according to the U.S. Salinity Laboratory, [From Richards (1954)].

8.13 EFFECT OF SALINIZATION AND SODICATION ON PLANT GROWTH

The accumulation of soluble salts in soils severly inhibits plant growth. It induces plasmolysis (see Sec. 5.4), by which H_2O moves out of the plant into the soil solution.

The presence of high amounts of Na^+ ions may keep the soil particles suspended. Upon drying, the soil may cake, and crust formation develops at the surface. The latter decreases soil porosity and aeration is severely inhibited.

The high pH in many of the soils also reduces availability of a number of micronutrients. These soils frequently encounter Fe, Cu, Zn, and/or Mn deficiencies.

8.14 IRRIGATION OF SALINE AND SODIC SOILS

Reclamation and management of the saline and sodic soils are based mainly on proper irrigation and drainage, on the exchange of Na^+ for Ca^{2+} on the exchange complex, and on the use of salt-tolerant crops.

Salinity Hazard

To make saline soils arable, leaching of excess salts by irrigation is usually conducted. A proper drainage method and the use of irrigation water with the proper salt quality are necessary. In this respect, the electrical conductivity EC_e is frequently used as an index for salinization hazard. The hazard of salinization is considered low if the irrigation water used has an $EC_e < 0.75$ mmho/cm (Richards, 1954; Taylor and Ashcroft, 1972).

Salinity hazard	EC_e, mmho/cm at 25°C
Low	<0.75
Medium	0.75-1.5
High	1.5-3.0
Very high	>3.0

In arid regions, salinization is a natural phenomenon. Therefore, the chances for salinization are considered very high if water with an $EC_e = 3.0$ or higher is used for irrigation over a period of many years, even on nonsaline soils.

Hazard of Sodication

The hazard of sodication is usually estimated by the use of the sodium adsorption ratio (SAR). The SAR formula is as follows (Richards, 1954):

$$SAR = \frac{(Na^+)}{\sqrt{(Ca + Mg)/2}} \quad (8.8)$$

The concentration of Na^+ and $Ca^{2+} + Mg^{2+}$ can be expressed in millimoles per liter (Kamphorst and Bolt, 1976) or in milliequivalents per liter (Taylor and Ashcroft, 1972). Since the sodic soils are highly saturated with Na^+ ions, it is necessary to use irrigation water with low SAR values on these soils. It is sometimes suggested to add gypsum to the irrigation water. But most often the gypsum is plowed under in the soil. This may ensure the development of low SAR values of the water in the soil. The Ca^{2+} may at the same time replace Na^+ from the exchange complcs. Theoretically, any soluble Ca compound, that will not affect soil pH, can be used together with irrigation water to reduce the SAR value and exchange the Na^+ ions.

8.15 SALT BALANCE AND LEACHING RATIO

Irrigation will sometimes only wet the soil. This is a potential danger for salt build-up. In the management of saline and sodic soils the so-called salt balance is taken into account. The latter means that the amount of salt brought into the soil must equal the amount of salt leached out of the soil. Therefore, more water must be applied over that needed to wet the soil. The additional water, used for leaching, is called the *leaching requirement* (LR) (Bernstein and Francois, 1973):

$$LR = \frac{EC_{iw}}{EC_{dw}}$$

where

EC_{iw} = electrical conductivity of irrigation water

EC_{dw} = electrical conductivity of saturation extract of saline soil which exhibited 50% decrease in yield

If the irrigation water increases in salinity over the years, the value of LR becomes larger. Bernstein and Francois (1973) suggested managing irrigation so as to supply the major water needs of crops at the minimum salinity level of the irrigation water.

8.16 IRRIGATION-INDUCED SALINIZATION AND SODICATION

The hazard of salinization and sodication is perhaps not limited to semiarid and arid region soils. Currently, it is common practice to also use supplementary irrigation in areas with measurable rainy seasons. With the easy access of water from the huge underground aquifers in the southern coastal plain region of the United States, large areas are now continuously being irrigated by the center pivot sprinkle system. No adequate disposal systems of the used irrigation water have been devised at present. The irrigation water reaching the soil is allowed to percolate naturally through the soil and return to the ground water in a more concentrated condition. A large part of the irrigation water may, perhaps, also evaporate leaving the salts in the surface soil. No investigations have been done yet as to the potential hazard in salinization and sodication by the use of this water. The danger of salinization and sodication is, of course, somewhat reduced by the presence of a humid climate. However, over a period of many years one can expect a reduction in quality of this irrigation water by the use of high amounts of fertilizers and in the absence of a proper drainage and disposal system.

9

SOIL CHEMISTRY AND SOIL FORMATION

9.1 WEATHERING PROCESSES

Weathering refers to the disintegration and alteration of rocks and minerals by physical and chemical processes. Physical weathering is caused by physical stresses within the rock or mineral. It causes the rocks to disintegrate into smaller-sized material without changing the chemical composition. Chemical weathering is caused by chemical reactions and definite chemical changes occur in the weathering products. For an illustration of the chemical reactions involved (e.g., solution, hydration, hydrolysis, oxidation, reduction, and carbonation) reference is made to the textbooks on principles of soil science (Brady, 1974; Donahue et al., 1977; Foth and Turk, 1978).

In nature both physical and chemical weathering may occur simultaneously. Both usually precede soil formation from solid rocks. Although by nature, physical weathering is of more importance at or near the soil surface, in some cases it may take place below the soil surface. Plant roots may contribute to physical weathering below the soil surface. By growing into cracks, they may rupture the rocks apart.

Chemical weathering can occur at the soil surface, in the solum or below the solum (in the parent material). For these reasons Jackson and Sherman (1953) suggested distinguishing it into pedochemical and geochemical weathering. *Pedochemical* weathering refers to chemical weathering within the solum, whereas *geochemical* weathering is weathering below the solum. Essentially, a sharp separation between pedochemical and geochemical weathering, as reported by Buol et al. (1973), is difficult to realize in nature. The main chemical reactions, such as solution, hydrolysis, hydration, oxidation, reduction, and carbonation, take place in the solum as well as in the parent material. Leaching of K from micas, alteration of clays by H^+ ions, and interlayering and formation of clays can occur as a pedochemical or as a geochemical process. Regardless of this difference in opinion, weathering in general results in a decrease in particle size of materials, in the release of soluble material, and in the synthesis of new materials (clays and humus).

9.2 STABILITY AND WEATHERING OF SOIL MINERALS

Crystal Chemistry and Mineral Properties

The breakdown and stability of minerals are quite complex and require a complete understanding of crystal chemistry. The relative resistance of a mineral to weathering processes is determined by its internal structure. The latter depends on the strength of the atoms or ions binding their neighboring ions in the crystal lattice of the mineral. Four major types of binding forces between atoms in crystals have been reported (Evans, 1939), i.e., ionic, homopolar, metallic, and Van der Waals forces. Although in many cases the structural bonds arising from interionic interactions cannot be attributed to one bond type, the bonds in the crystal structure of soil minerals are considered to be mostly ionic in nature.

A number of mineral properties are affected by the respective bond types (Table 9.1). The ionic and homopolar bonds between atoms yield hard crystals with high melting points. On the other hand, Van der Waals attraction gives rise only to weak bonds and relatively soft crystals with low melting points.

Coordination Theory and Pauling's Rules

The structure of soil minerals is formed by regular groupings of anions packed closely around a cation. Since most soil minerals are in oxide forms, the anions are usually oxygen atoms. The number of anions surrounding the cation is called the coordination number and is dependent upon the respective ionic radii. A cation surrounded by three anions in an equilateral triangular configuration has a coordination number of 3. An arrangement of four anions around the cation is called a fourfold coordination. Such a crystal structure is called a *tetrahedron*. Silica tetrahedrons are examples of fourfold coordination structures. A sixfold coordination structure is a configuration with six anions around the cation, yielding an octahedron, such as the aluminum octahedron. With ions of larger dimensions, coordination numbers of 7 to 12 are possible.

This coordination theory in crystal structure is slightly different than the one used in complex compounds (chelates) formed by coordination reactions. However, viewed from the standpoint of broad generalities some similarities between the two types of coordination theories are present. Complex formation and chelation by coordination reactions will be discussed in Chapter 10.

Since the atomic bonds in many of the soil minerals are ionic in nature, the crystal structure of these minerals obeys the principles of ionic crystals as formulated by Pauling's rules (1929). The first rule states that "a coordinated polyhedron of anions is formed about each cation with the cation-anion distance being determined by the sum of

Table 9.1 Selected Physical and Structural Properties of Minerals as Related to Bonding Type

Mineral property	Type of Bonds			
	Ionic	Homopolar	Metallic	Van der Waals
Mechanical	Strong, hard	Strong, hard	Variable	Weak, Soft
Thermal	High melting point	High melting point	Variable melting point	Low melting point
	Low thermal expansion	Low thermal expansion	–	High thermal expansion
Electrical	Nonconducting	Nonconducting	Conducting	Nonconducting
Optical	Variable	High refractive index	Opaque	Transparent
Structural	High coordination	Low coordination	Very high coordination	Very high coordination
	Moderately high density	Low density	High density	–

Source: Evans (1939)

the radii, and the coordination number of the cation being determined by the radius ratio." Not only are crystal structure and associated coordination number dependent on the size of the ions, they must at the same time obey the law of electroneutrality. A large cation, therefore, can coordinate more anions around it and still keep the anions apart. Smaller cations are capable of coordinating a smaller number of anions. The limiting factor for each crystal arrangement is formulated by the radius ratio of the ions involved. The radius ratio is expressed as follows:

$$\text{Radius ratio} = \frac{r_c}{r_a}$$

where

r_c = radius of cation

r_a = radius of anion

A minimum radius ratio exists for each coordination number (Table 9.2). If the radius ratio is below a minimum value, the cation can only coordinate the next smaller number of anions. For example, a limit range of the radius ratio of 0.155 to 0.255 indicates that the cation in any cation-anion combination with a radius ratio r_c/r_a of 0.155 is capable of being in close association with only three anions. It will coordinate four anions if the radius ratio satisfies the value of 0.255 or larger.

It should be noted again that the values in Table 9.2 are the ranges within which the radius ratio of atoms arranged geometrically in a crystal can vary without affecting the corresponding coordination number. The absolute values of r_c/r_a ratios between any two atom pairs are generally smaller than those stated in Table 9.2. For example, the absolute radius ratio of silicon and oxygen r_{Si}/r_O = 0.42/1.40 = 0.300 qualifies for a fourfold coordination.

Table 9.2 Relationship of Radius Ratio r_c/r_a and Coordination Number

Limit range of r_c/r_a	Coordination number	Crystal geometry
0–∿	1	
0–∿	2	Angular
0.155–0.255	3	Trigonal planar
0.255–0.414	4	Tetrahedron
0.414–0.732	6	Octahedron
0.732–1.0	8	

However, the absolute radius ratio between aluminum and oxygen is $r_{Al}/r_O = 0.51/1.40 = 0.364$. The latter does not fall within the limit range for a sixfold coordination.

Pauling's second rule states that "in a stable coordination structure, the total strength of the valency bonds which reach an anion from all neighboring cations equals the charge of the anion." This rule, also known as the *electrostatic valency principle*, indicates that the charge of the cation must be shared (divided) equally by the number of bonds to the neighboring anions. At the same time, the number of these bonds depends upon the coordination number of the cation. This rule provides for a symmetrical arrangement of bonds of equal strength around each cation. Applied to soil minerals, Pauling rule results in the following arrangement of charges. In soil minerals, mostly oxides, the number of oxygen atoms normally packed around the cation is considered the coordination number. The coordination number of silicon is 4, that of aluminum may be either 4 or 6, while iron and magnesium have coordination numbers of 6. Potassium is a large ion and is characterized by a coordination number of 12. Since each silicon ion in the tetrahedron is surrounded by four oxygen atoms, Pauling rule says that the bond strength is equally shared among the tetrahedral bonds; in other words, the charge of the silicon ion is divided by the number of bonds (by 4). Consequently, each oxygen atom in the tetrahedron has half of its charge satisfied by the silicon to which it is bonded (see Figure 9.1). The remaining unsatisfied valencies of the oxygen atoms are balanced by association with another silicon or with two Al^{3+} and/or Mg^{2+} ions. The sharing of an oxygen atom by two silicon ions develops the basic silicate mineral structure.

In an aluminum octahedron, the Al^{3+} ion is surrounded by six oxygen atoms. The bond strength with each oxygen atom contributed by the Al^{3+} ion is, therefore, 3/6 or 1/2. Applying Pauling's second rule in a similar manner, Fe^{2+} or Mg^{2+} in octahedral position will contribute only 2/6 = 1/3 charge to each of the oxygen atoms.

Stability of Minerals and Bond Strength

As discussed previously the fundamental units of silicate minerals are SiO_4 tetrahedrons. The latter can be joined together in several ways, and depending on the arrangement of these SiO_4 tetrahedra, the minerals have been distinguished into cyclo-, ino-, neso-, phyllo-, soro-, and tectosilicates (Sec. 4.3). Single or several units of tetrahedra can be linked together by other cations in the mineral framework. For example, double chains of silica tetrahedra can be linked together by Ca and Mg atoms, such as in amphiboles; or SiO_4 and AlO_4 tetrahedra are linked together by alkali and/or alkaline earth metals located in the lattice interstices, such as in feldspars. Whatever the structural arrangement is, the Si-O-Si linkage, called the *siloxane linkage*

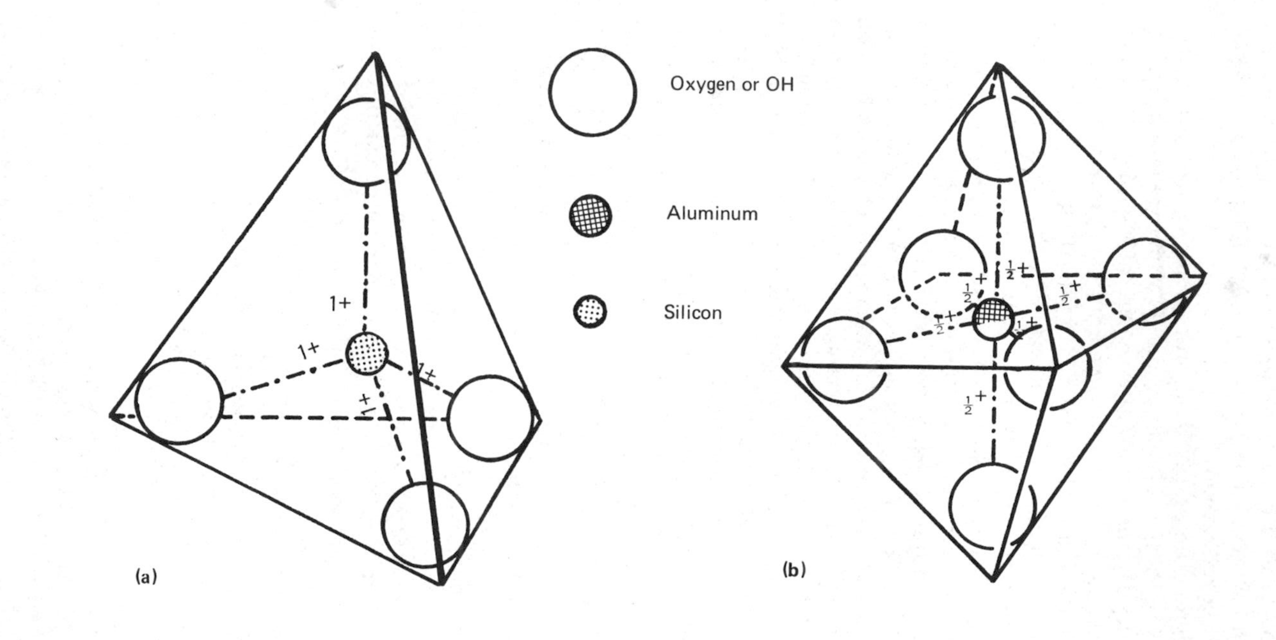

Figure 9.1 (a) The charge of the silicon atom is shared equally with the four surrounding oxygen atoms. Consequently each oxygen has only half of its charge satisfied by the silicon, and (b) the charge of the aluminum atom is shared equally with the six surrounding oxygen atoms. Therefore, the bond strength contributed by the Al^{3+} ion to each oxygen atom is only 1/2.

(Sticher and Bach, 1966), requires the greatest energy to form, compared to the other cation-oxygen bonds (Table 9.3). The data in Table 9.3 show that Si-O bonds are the strongest bonds, requiring 3110 to 3142 kg cal/mol for their formation. Aluminum-oxygen bonds are the next strongest (1793 to 1878 kg cal/mol needed for formation), while the bonds between the other metal ions and oxygen appear to be the weakest (299 to 919 kg cal/mol). The greater the number of Si-O bonds by linkage of increasingly larger numbers of silica tetrahedra through oxygen sharing, the greater will become the resistance to weathering. On the basis of a progressive increase of oxygen sharing between adjacent silica tetrahedra, Keller (1954) and Birkeland (1974) ranked the silicate groups as follows: Nesosilicates < inosilicates (single chain) < inosilicates (double chain) < phyllosilicates < tectosilicates.

Table 9.3 Energies of Formation of Cation-Oxygen Bonds

Cation	Energy of formation, kg cal/mol
Si^{4+} (Nesosilicates)	3142
Si^{4+} (Inosilicates, single chain)	3131
Si^{4+} (Inosilicates, double chain)	3127
Si^{4+} (Phyllosilicates)	3123
Si^{4+} (Tektosilicates)	3110
Al^{3+} (framework)	1878
Al^{3+} (non-framework)	1793
Fe^{3+}	919
Mg^{2+}	912
Ca^{2+}	839
H^{+} (in OH)	515
Na^{+}	322
K^{+}	299

Source: Paton (1978) and Keller (1954).

To correlate such a ranking with a corresponding increase of resistance against weathering is subject to question. For example, the clay minerals, belonging to the phyllosilicates, are more resistant to weathering than feldspar and leucite, which are tectosilicates. However, in terms of comparison between biotite (phyllosilicates) and feldspar (tectosilicate), the ranking above may have some value. Apparently, susceptibility to weathering is not affected only by the mineral structure, but a number of additional factors may also play an important role in mineral breakdown, as will be discussed with the individual silicate groups.

If the assumption above is valid, stating that bonds requiring the greatest energy to form will also be the most resistant to attack by weathering, the data in Table 9.3 suggest that destruction of nonframework cation-oxygen bonds, such as Mg-O and Ca-O, will be relatively easier than the decomposition of the siloxane bonds, considered to be the silicate framework bonds. Cleavage of the siloxane bonds is made possible especially by interaction with chelating substances (Sticher and Bach, 1966). After collapse of the nonframework bonds, the tetrahedra may begin to break down, where aluminum is present in tetrahedral position. Since the weaker bonds are subject first to attack, the implication is that energy requirements for weathering may be considerably less than energy requirements for formation of the bonds.

In view of the above consideration, the minerals in the various silicate groups are expected to differ in the way they respond to attack by weathering.

Cyclosilicates The structure of this group is characterized by six-membered hexagonal rings of silica tetrahedra linked together by cations, such as Mg, Na, and/or Fe. The bonds formed by the latter are the weakest spots, but because of the abundance of Si-O linkages, the minerals in this group are considered relatively stable.

Inosilicates This group has in its structure single-chain (pyroxenes) and double-chain (amphiboles) silica tetrahedra linked together by Ca, Mg, and/or Fe. Due to the presence of many weak spots provided by the Ca-O, Mg-O, and/or Fe-O bonds, these minerals tend to weather rapidly.

Nesosilicates The minerals in this group are composed of single tetrahedra linked together by Mg^{2+} and Fe^{2+} ions. To effect a breakdown, it is considered sufficient to sever the weaker Mg-O and/or Fe-O bonds. Notwithstanding bond energy considerations, susceptibility of the minerals in this group to breakdown by weathering appears to vary considerably from one mineral to another, e.g., olivine versus zircon. The tight packing of oxygen atoms known to exist in zircon makes this mineral comparatively hard. On the other hand, the looser packing of oxygens in olivine makes the mineral weather faster.

Phyllosilicates Linkages of silica tetrahedra and aluminum octahedra sheets by mutually shared oxygen atoms form the basis for the structure of this group. Some of the minerals, e.g., biotite and muscovite, are relatively susceptible to weathering; others like the clay minerals are resistant weathering products and further breakdown of clays is difficult. Disruption of the mineral usually occurs through removal (or replacement with OH) of interlayer ions or through cleavage of Al-O bonds in tetrahedral and octahedral positions.

Sorosilicates Individual and linked silica tetrahedra formed by mutually shared oxygen are the basis for the structure of this group. Consequently they are rather difficult to decompose. However, decomposition may take place in tetrahedrons where Al has substituted for Si.

Tektosilicates The minerals are considered solid solution minerals with a framework of silica tetrahedra, in which the cavities are occupied by Na, Ca, and so on. The minerals in this group may also vary considerably in their resistance to weathering, e.g., leucite and plagioclase versus potash feldspars. The relative degree of close packing of atoms in their structural framework may be the reason for such a variability in weathering. Increased substitution of Al for Si in tetrahedra of plagioclase minerals is also considered a factor that makes this mineral weaker than potash feldspar.

9.3 WEATHERING OF FELDSPARS AND THE SILICA POTENTIAL

An important process in chemical weathering is the decomposition of soil minerals by hydrolysis. This can be illustrated by the decomposition reaction of orthoclase:

$$2KAlSi_3O_8(c) + 2H^+(aq) + 9H_2O(l) \rightleftharpoons H_4Al_2Si_2O_9(c) + 4H_4SiO_4(aq) + 2K^+(aq) \qquad (9.1)$$

By assuming the activities of orthoclase, water, and kaolinite unity, the mass action law gives

$$K = \frac{(H_4SiO_4)^2(K^+)}{(H^+)}$$

in which K denotes the decompositon or equilibrium constant. By taking the log, the equation above becomes

$$\log K = \log \frac{(K^+)}{(H^+)} + 2 \log (H_4SiO_4) \tag{9.2}$$

In p (-log) form, Eq. (9.2) is considered the chemical potential determining weathering stability of orthoclase. If the activity of H_4SiO_4 decreases below the value of the silica potential (pH_4SiO_4) of quartz or amorphous silica, orthoclase will decompose and form kaolinite as indicated by reaction (9.1). In equilibrium condition the ratio $(K^+)/(H^+)$ is unity (Garrels and Christ, 1965), and only the activity of H_4SiO_4 remains to control stability of orthoclase.

9.4 WEATHERING OF KAOLINITE AND THE GIBBSITE POTENTIAL

After formation, kaolinite is also subject to decomposition and can be transformed into gibbsite by weathering. The following reaction (Kittrick, 1967, 1969; Tan et al., 1973) determines the stability of kaolinite:

$$Al_2Si_2O_5(OH)_4(c) + 5H_2O(aq) \rightleftharpoons 2Al(OH)_3(c) + 2H_4SiO_4(aq) \tag{9.3}$$

In pure condition, kaolinite, gibbsite, and H_2O are considered unity, therefore the following relationship follows from reaction (9.3):

$$pK = 2\ pH_4SiO_4 \tag{9.4}$$

The assumption is made that after part of the kaolinite is converted into gibbsite, the solution soon becomes saturated relative to gibbsite. At this point, kaolinite and gibbsite exist together in equilibrium, satisfying reaction (9.3). Equation (9.4) is considered the chemical potential determining stability of kaolinite. Kaolinite is stable if pH_4SiO_4 is smaller than 4.73 (Tan et al., 1973).

Gibbsite may in turn be converted into Al^{3+}. The decomposition reaction can be written as

$$Al(OH)_3(c) + 3H^+(aq) \rightleftharpoons Al^{3+}(aq) + 3\ H_2O(l)$$

for which is valid

$$pH - \frac{1}{3} pAl^{3+} = 2.7 \tag{9.5}$$

The expression pH - 1/3 pAl^{3+} is called the *gibbsite potential*. Again, it can be argued that this chemical potential determines stability of gibbsite. Gibbsite is stable only if the gibbsite potential is less than 2.7 and when the activity of H_4SiO_4 is very small (Kittrick, 1967, 1969; Tan et al., 1973).

9.5 STABILITY AND PHASE RELATIONSHIPS OF SOIL MINERALS

After the stability of feldspar, kaolinite, gibbsite, and/or other soil minerals have been formulated as chemical potentials, a phase diagram can be drawn. This diagram delineates the regions where the minerals are stable and shows the points where they are not stable and start to decompose to form another mineral. An example of such a diagram is given in Figure 9.2.

The line bordering the areas of gibbsite and kaolinite is the chemical potential as expressed by $pH_4SiO_4 = 4.73$. All points on this line represent conditions at which both kaolinite and gibbsite coexist in equilibrium. In the gibbsite area (to the left of the line, $pH_4SiO_4 > 4.73$), only gibbsite is stable. If kaolinite is present at $pH_4SiO_4 > 4.73$, it will automatically decompose and form gibbsite. In

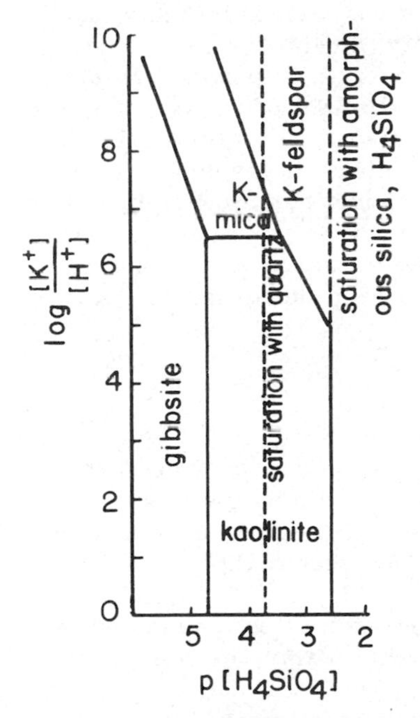

Figure 9.2 Stability and phase diagram of soil minerals.

the kaolinite area (to the right of the line, $pH_4SiO_4 < 4.73$), kaolinite is stable. If gibbsite is present at $pH_4SiO_4 < 4.73$, the presence of H_4SiO_4 will resilicate gibbsite into kaolinite. A similar discussion can be given for the other lines and areas.

Many types of stability diagrams can be made. Some are relatively simple, others are three dimensional and very complex (Garrels and Christ, 1965). However, all of them have as a purpose the prediction of possible successive alteration of minerals with gradual changes in ion activity ratios as weathering proceeds.

9.6 EFFECT OF CHELATING AGENTS ON WEATHERING

Evidence has been reported that soil organic matter has a significant effect on weathering. In few cases the degree of weathering induced by soil organic matter may be more important then that brought about by chemical reactions alone.

By the decomposition of organic matter, a number of organic compounds are released or synthesized. Most of them, such as humic and fulvic acids, have the capacity to chelate or complex metal ions (Schnitzer and Kodama, 1976; Tan, 1978a, 1976a). Therefore, they may be able to pry loose Al and Fe from micas, feldspars, and kaolinite or any other soil mineral, accelerating in this way the decomposition process. The organic chelating agent may perhaps react with an exposed cation, followed by the movement of the complex compound or chelate into solution. As a chelate, Al and other metals may be rendered soluble over a pH range in which it is insoluble as an ion. This is of importance in the formation of spodic horizons in Spodosols. Also, the H^+ ions produced during decomposition of organic matter may be adsorbed by silicate clays. H^+ saturation of clays which results in their gradual decomposition has been discussed earlier.

9.7 SOIL FORMATION PROCESSES

The process of soil formation is a complex biological and chemical problem and is usually difficult to describe with a single reaction. Reactions may occur simultaneously, or a sequence of reactions one after another are involved. Simonson (1959) stated that the soil pedon is formed by the combined effort of additions of inorganic and organic materials to the surface, transformation of compounds within the soil, vertical transfer of soil constituents within the soil, and removal of soil components from the soil.

The types of processes involved vary according to the conditions. However, it is not the purpose of this topic to list them in this chapter. For a list of possible processes of soil formation reference is made to Buol et al. (1973).

It is perhaps more important to discuss major soil forming processes with general applicability to development of soil pedons, such as desilication and translocation of Al and Fe related in the formation of argillic, albic, spodic, and oxic horizons.

Desilication

This is a process in which silica is released from soil silicates. Part of the released silica reacts with alumina to form clays, whereas the remainder is subject to leaching. Consequently the soil exhibits a loss in silica content, and has at the same time a residual accumulation of stable weathering products, including sesquioxides.

A process such as desilication may occur in the tropics or in the temperate regions in the presence of sufficient amounts of moisture and the right temperature. Usually it is more pronounced in the humid tropics. It is known in the past as laterization or ferrallitization. The reactions can be illustrated with the decomposition of orthoclase into kaolinite [Eq. (9.1)] and with the decomposition of kaolinite into gibbsite [Eq. (9.3)]. If the soil is well drained and permeability is rapid, the activities of dissolved ions and H_4SiO_4 are kept low by leaching. The end product of weathering will then be gibbsite. Under poorly drained conditions and slow permeability, leaching is inhibited. The latter results in an increase in H_4SiO_4 activity and $(K^+)/(H^+)$ ratio, leading to formation of illites (Van Schuylenborgh, 1971).

The degree to which Si can be leached out of soil depends on its capacity to remain in solution. The solubility of silica is determined by the law of polymerization. Present in concentrations below 140 ppm SiO_2 (25°C), silica is found mainly in the form of monosilicic acid, $Si(OH)_4$, which is considered a true solution (Millot, 1970; Krauskopf, 1956). The solubility of this silica remains constant at 140 mg/l in the range of pH 2 to 9 (see Figure 9.3), but at pH values above 9, solubility of silica increases rapidly. If the concentration of silica in the solution exceeds 140 ppm, polymerization of silica occurs, and usually a mixture of polymers and monomers of $Si(OH)_4$ is found in the soil solution. The polymers will be precipitated by the introduction of small quantities of Al, or by decreasing pH (Paton, 1978), leaving the monomers in solution which may tend to be leached out of the system.

Translocation of Clays

This process leads to enrichment of B horizons with clays. Such B horizons are called *argillic horizons* in the U.S. Soil Taxonomy. The clays have migrated from the A horizon because of an increase in peptization.

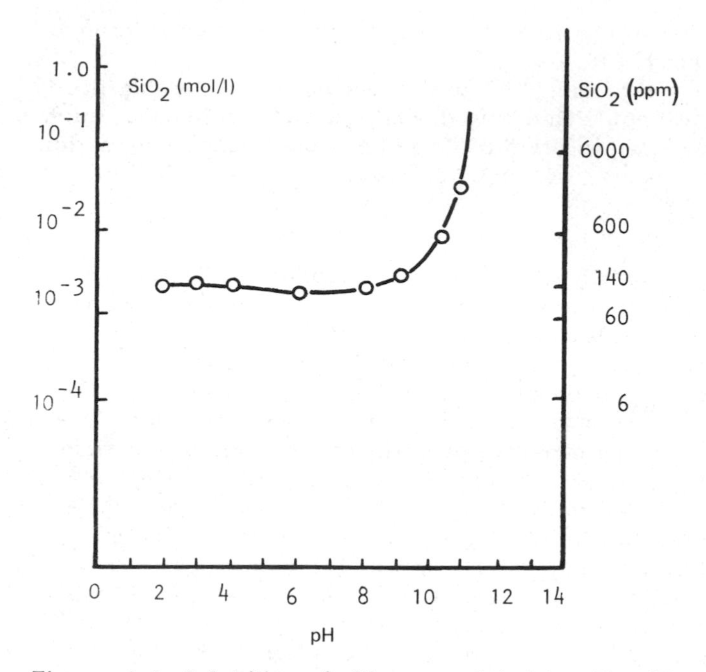

Figure 9.3 Solubility of silica as related to pH. [From Krauskopf (1956).]

Evidence has been presented that clays and organic matter can form complexes (Greenland, 1971; Tan, 1976a). Although the exact mechanism is not known yet, the following hypothetical reaction serves as an example:

$$\begin{matrix} -Si-O \\ | \quad\quad \rangle Al-OH \\ -Si-O \end{matrix} + HO-C_6H_4-COOH \longrightarrow \begin{matrix} -Si-O \\ | \quad\quad \rangle Al-O-C_6H_4-COOH \\ -Si-O \end{matrix} + H_2O$$

CLAY — Org. comp. — Clay-organic complex

The reaction above adds to the clay surface an acidic group (COOH), and contributes to the clay a strong negative charge. The surface potential of the clay-organic complex is, therefore, larger than that of the clay alone. Consequently the electrokinetic potential, formerly explained with the ζ potential, also becomes larger. This increases the capacity for peptization of clays. As a clay-organic complex, the clay remains suspended for a longer time and moves downwards with the percolating water.

A number of reactions are responsible for clay accumulation in the B horizon. The movement stops where the percolating water stops and flocculation of clay may occur. Capillary withdrawal of water into the soil fabric deposits clay as clay skins on the walls of pores and peds.

Translocation of Al and Fe

The downward movement of Al and Fe together with organic matter results in the formation of albic (A_2) and spodic (B_{hir}) horizons. This process was called *podzolization* in the past. It gives rise to the development of Spodosols (Podzols).

Most of the Fe subject to translocation comes from the decomposition of biotite and ferromagnesian minerals. The possible ionic forms of Fe(III) are Fe^{3+}, $(FeOH)^{2+}$, $Fe(OH)_2^+$, $Fe_2(OH)^{4+}$ and $Fe(OH)_4^-$ (Van Schuylenborgh, 1966). The ionic forms of Fe(II), e.g., Fe^{2+}, $(FeOH)^+$, and $Fe(OH)_3^-$ are less stable than those of Fe(III). Most of the soils, where translocation occurs, are well drained. Therefore, most of the iron is in Fe(III) ionic form. The concentration of the Fe ions depends on the solubility of their respective solid phase.

Solubility constants of Fe compounds (Van Schuylenborgh, 1966)

$Fe(OH)_3$ amorphous $\rightleftharpoons Fe^{3+} + 3OH^-$	$pK = pFe + 3pOH = 38.2$
$Fe(OH)_3$ amorphous $\rightleftharpoons (FeOH)^{2+} + 2OH^-$	$pK = p(FeOH) + 2pOH = 26.3$
$Fe(OH)_3$ amorphous $\rightleftharpoons Fe(OH)_2^+ + OH^-$	$pK = pFe(OH)_2 + pOH = 17.0$
Goethite	pK = 45.2
Lepidocrocite	pK = 42.5
Hematite	pK = 42.5
Maghemite	pK = 41.0

The assumption was made above that in well-drained soils the iron is in Fe(III) form. However, whether Fe(III) or Fe (II) occurs under natural soil condition depends more precisely on the oxidation potential. If the following redox reaction of iron is studied:

$$Fe^{2+} \rightleftharpoons Fe^{3+} + e^-$$

$$\text{then } K = \frac{[Fe^{3+}]}{[Fe^{2+}]}$$

Application of the Nernst equation gives the following relation:

$$E_h = E° + \frac{RT}{nF} \ln K$$

$$\text{or} \quad E_h = E° + 0.059 \log \frac{[Fe^{3+}]}{[Fe^{2+}]}$$

in which E_h is called the *oxidation potential*. After complete oxidation of Fe(II) to Fe(III), the concentration of Fe(II) ions becomes negligible small and can be neglected, so that the relationship changes to

$$E_h = E° + 0.059 \log [Fe^{3+}]$$

If, however, reduction processes prevail, the activity of Fe(II) becomes very large, so that for all practical purposes, activity of Fe(III) can be neglected. The oxidation potential assumes then the following relation:

$$E_h = E° - 0.059 \log \frac{[Fe^{2+}]}{[Fe^{3+}]}$$

Fe^{3+} neglibible:

$$E_h = E° - 0.059 \log [Fe^{2+}]$$

Therefore, the oxidation potential E_h increases upon oxidation and decreases as a result of reduction processes. When the activity of Fe(III) equals that of Fe(II), the oxidation potential also equals the standard oxidation potential:

$$[Fe^{3+}] = [Fe^{2+}] \qquad E_h = E°$$

Using these oxidation potentials and pH values, a number of stability fields of Fe(II) and Fe(III) systems have been developed by various authors (Garrels and Christ, 1965; Hem and Cropper, 1959). An example is give in Figure 9.4.

From the diagram in Figure 9.4, it could be noticed that Fe(II) ions are stable at oxidation potentials below $E_h = 0.3$ if soil pH = 5 to 7. Only if the soil reaction is strongly acid, will Fe(II) ions remain stable at $E_h > 0.3$. The natural condition is represented by the shaded portion in Figure 9.4. A major part of this shaded area (2/3) lies in the stable field of ferric hydroxide, $Fe(OH)_2^+$, compounds. Hem and Cropper (1959) indicated that Fe(II) ions remain soluble at concentrations not to exceed 100 ppm ($E_h = 0.3$ and pH = 5.0). For a further discussion and detailed treatise on the principles of redox potentials see Chap. 2 and Sec. 9.8.

Almost all silicates are sources for Al. The ionic forms of Al(III) are Al^{3+}, $(AlOH)^2$, $Al(OH)_2^+$, $Al(OH)_4^-$, $Al_2(OH)_2^{4+}$, $Al_2(OH)_4^{2+}$, $Al_4(OH)_{10}^{2+}$, and $Al_6(OH)_{12}^{6+}$ (Van Schuylenborgh, 1966). As is the case with iron, the concentration of the various Al ions is dependent on the solubility of their respective solid forms.

Solubility constants of Al compounds (Van Schuylenborgh, 1966)

$Al(OH)_3$ amorphous $\rightleftharpoons$ Al^{3+} + $3OH^-$	pK = pAl + 3pOH = 32.0
$Al(OH)_3$ amorphous $\rightleftharpoons$ $(AlOH)^{2+}$ + $2OH^-$	pK = pAlOH + 2pOH = 23.4
$Al(OH)_3$ amorphous $\rightleftharpoons$ $Al(OH)_2^+$ + OH^-	pK = $pAl(OH)_2$ + pOH = 14.1
Gibbsite	pK = 36.2 (Al^{3+})
Gibbsite (in H_2O)	pK = 14.6 ($Al[OH]_4^-$)
Gibbsite (in base)	pK = 0.57 ($Al[OH]_4^-$)

The data above show that most of the solubility constants (pK values) of Al and Fe compounds are large. This means that the solubility of the compounds are very low. The lower the pK value, the more soluble the compounds are in soils. It appears that only gibbsite in alkaline condition has a low pK value, perhaps sufficiently low to release some Al ions in solution. In addition to this, one has also to take into consideration that the pH range in many soils is such that most Fe and Al compounds are essentially insoluble. Therefore, the possibility of migration of Al and Fe in the ionic forms above is very small. Other agents are required to make Fe and Al more soluble.

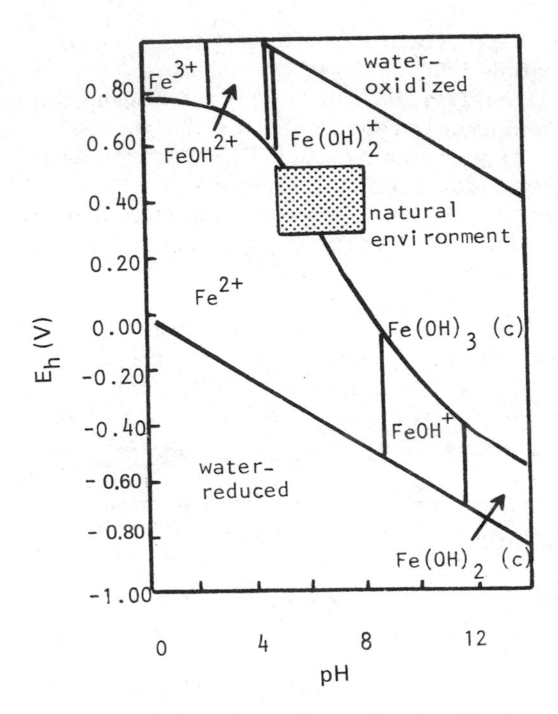

Figure 9.4 Stability field diagram for aqueous Fe(III)-Fe(II) systems. [From Hem and Cropper (1959).]

Evidence has been presented showing that decomposition products of soil organic matter (Hodgson, 1969; Martin and Reeve, 1960a, 1960b; Tan, 1978a) are capable of forming complexes with Fe and Al. As a complex, Fe and Al may remain soluble at pH ranges that make them usually insoluble. Another possibility is that the metal-organic complexes tend to disperse easily at low electrolyte concentration. In this way, they may move downward to deeper layers in the soil pedon. The loss of Fe, Al, and organic matter from the A horizon leads to the formation of an albic horizon.

Since the solubility constants of Al and Al-organic complexes are smaller than those of Fe and Fe-organic complexes, the Al complexes are more soluble than the Fe complexes. Consequently Al complexes may move deeper in the soil pedon than the Fe complexes. In the B horizon, these compounds will be accumulated by (1) formation of insoluble complexes, (2) hydrolysis of metal complexes, (3) microbial attack of the organic ligands (Van Schuylenborgh, 1965), or (4) by a combination of these processes.

9.8 OXIDATION AND REDUCTION REACTIONS IN SOILS

Reduction and oxidation reactions occur almost in any soils. Reduction is by definition the gain of electrons, whereas oxidation is the loss of electrons. This can be illustrated by the following reaction:

$$Fe^{3+} + e^{-} \rightleftharpoons Fe^{2+} \qquad (9.6)$$

Oxidation reactions are usually related to well-drained soil conditions. On the other hand, reduction processes are associated with poorly drained conditions, or where excess water is present. The latter developed gley formations.

Usually known as the *soil redox state,* it occurs almost in any soil. Both reduction and oxidation conditions can occur simultaneously in the pedon. While the surface layers of the pedon are in an oxidized state, the subsoil layers may be in a reduced condition due to a fluctuating ground water level. The latter may lead to pseudogley formation or to plinthization.

The redox condition of soils affect stability of iron and manganese compounds. Microbial activity, accumulation and decomposition of organic matter are to a certain extent also affected by the soils redox state. Fresh organic matter is thought to aid formation of reduced condition. Bloomfield (1951; 1953) reported that aqueous leaf extract reduced Fe(III) into Fe(II) in soils. In tidal flood water areas reduction processes play a considerable role in the formation of sulfur-rich soils.

Soils with different redox conditions may react differently upon N fertilization. In well-drained soils, ammonium N is subject to nitrification. However, if the ammonium fertilizer is applied to a reduced soil, such as to lowland rice or paddy soils, it remains available as ammonium.

Redox Potentials

The half-cell reaction for an oxidation-reduction system can be illustrated with reaction (9.6) and attains the following general expression:

$$\text{Oxidized state} + e \rightleftharpoons \text{reduced state}$$

The corresponding half-cell potential for the reaction above obeys the Nernst equation:

$$E_h = E_o + \frac{RT}{nF} \log \frac{(\text{oxidized state})}{(\text{reduced state})} \qquad (9.7)$$

E_h is the redox potential; it is in fact the half-cell potential relative to a standard reference electrode. E_o is a constant, called the *standard redox potential* of the system, and RT/F = 0.0592 at 25°C (see page 34; Garrels and Christ, 1965). If the activities of the oxidized and reduced species are unity, the ratio becomes 1, and the log equals 0. Consequently $E_h = E_o$. Therefore, the standard redox potential is defined as the redox potential of the system at which the activities of oxidized and reduced species are unity.

Application of Redox Potentials in Soils

The redox potential of soils varies with the reduced and oxidation state in soils. It is also associated with soil pH. E_h - pH relationships are usually linear in character (Garrels and Christ, 1965).

An illustration for the variations in redox potentials can be given when reduced iron is oxidized by adding an oxidizing agent (see Figure 9.5). The curve in Figure 9.5 indicates that oxidation of Fe(II) causes the redox potential to rise. When 50% of Fe(III) is present, the redox potential equals 770 mV. From the above, and from Eq. (8.6), it can be inferred that the redox potential of soils in an oxidized state is higher than that of soils in a reduced state. Jeffery (1960) reported an E_h value of -250 mV for soils in strong anaerobic condition. He also found, that the redox potential was affected by flooding as shown in Figure 9.6. During the initial stage of flooding the redox potential dropped rapidly, then it increased again and stabilized at approximately 100 mV. A system which has a stabilized redox potential is said to be *well poised.*

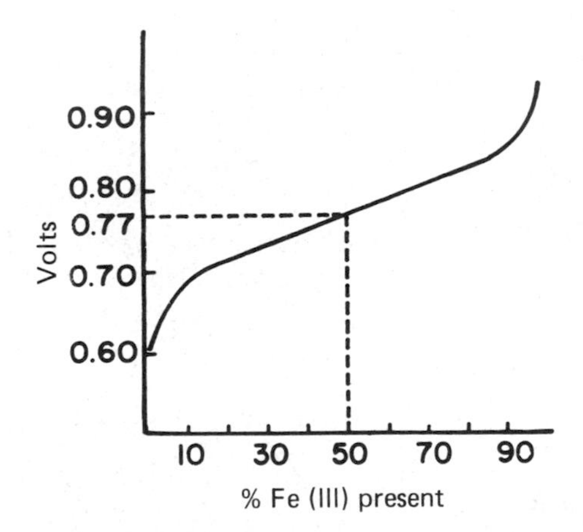

Figure 9.5 Redox potential curve of a waterlogged soil in the presence of oxidation of Fe(II) → Fe(III). [From Jeffery (1960).]

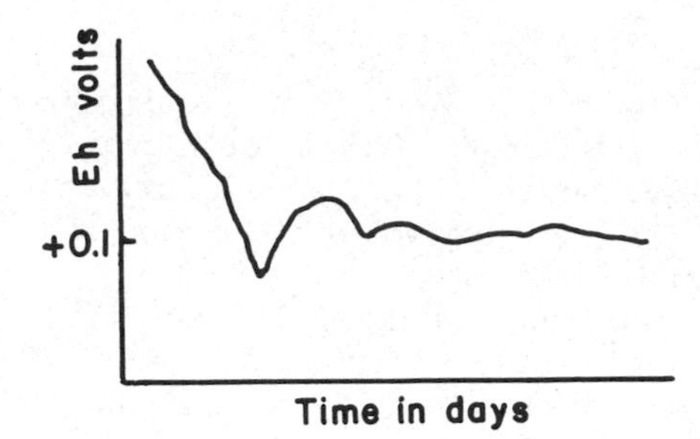

Figure 9.6 The redox potential curve as influenced by the length of flooding time. [From J.W. Jeffery (1960).]

Stability of Iron Oxide and Hydroxides

The redox potential and pH are used to define stability relationships between iron oxides and iron hydroxides minerals. The formation of hematite from magnetite is considered an oxidation reaction. The reaction is simplified as follows:

$$2Fe_3O_4 + H_2O \rightleftharpoons 3Fe_2O_3 + 2H^+ + 2e^-$$

The redox potential of the system above is then

$$E_h = E_o + \frac{0.059}{2} \log \frac{(Fe_2O_3)^3 (H^+)^2}{(Fe_3O_4)^2 (H_2O)}$$

By assuming that H_2O and the mineral species are in a pure state, their activities are unity. Therefore, the equation above changes to

$$E_h = E_o + \frac{0.059}{2} \log (H^+)^2$$

or $E_h = E_o - 0.059 \text{ pH}$ (9.8)

Equation (9.8) is a linear relationship between E_h and pH. It indiccates the boundary between stability of magnetite and hematite. If the redox potential of the system is larger than E_h of Eq. (9.8), hematite is stable. If the redox potential of the system is smaller than the E_h formulated by Eq. (9.8), magnetite exists as the stable species.

Activity of Reduction Products

Van Breemen and Brinkman (1976) stated that flooding of aerobic soils reduced first the NO_3^- in soils. After the disappearance of nitrate, manganese would be reduced, followed by iron. The latter increased the concentrations of Mn^{2+} and Fe^{2+} ions during the initial period of reduced condition. The concentrations of Mn^{2+} and Fe^{2+} ions decreased again upon continued flooding and stabilized at a constant level. The net reaction is a condition in which Fe(III) and Fe(II) ions are present together. Such a condition is considered desirable for soils (Jeffery, 1960).

10

CHEMISTRY OF SOIL-ORGANIC MATTER INTERACTION

10.1 COMPLEX FORMATION AND CHELATION

The terms *complex formation* and *chelation* have been used in soil science interchangeably. However, based on the nature of bonding a distinction can perhaps be made between complex formation and chelation. Complex formation is the reaction of a metal ion and ligands, through electron pair sharing (Murmann, 1964; Mellor, 1964; Martell and Calvin, 1952). The resulting product is called a *metal coordination compound.* The metal ion is the electron-pair acceptor, and the ligand is the electron-pair donor. The metal ion serves as the central ion, while the organic ions are coordinated around it in a first coordination sphere.

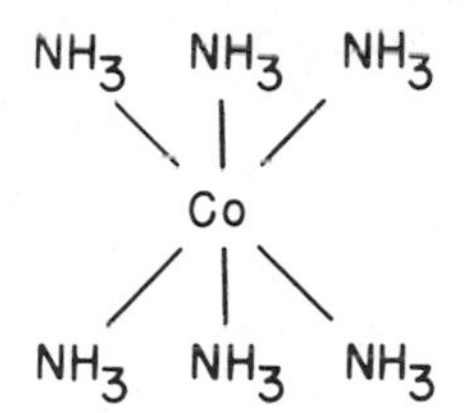

The number of ligands bonded to the central atom in a definite geometry is called the *coordination number.* Some of the organic ligands can bind the metal ion with more than one donor functional group. This type of bonding forms a heterocyclic ring, called a *chelate ring.* The process of formation of a chelate ring is called *chelation* (chelate means lobster claw). Hydrated metal ions in solution are also considered complexes with water since they are surrounded by H_2O molecules (Perrin, 1964).

Almost any metal atom can serve as an acceptor atom, including K^+, Li^+, Na^+, Ag^+, and Au^+ (Murmann, 1964). A long-known complex compound with a monovalent ion is potassium ferrocyanide. A ligand can be an anion (Cl^- or $R-CH_2-COO^-$) or a neutral molecule (NH_3). The complexes produced can be cations, anions, or neutral molecules.

For a review on the binding of metal ions by and the complexing capacity of humic acids, refer to Stevenson (1976a).

10.2 METAL-ORGANIC COMPLEX REACTIONS

Stability Constants

A number of organic compounds including humic and fulvic acids (see pages 48 to 84) are capable of forming complexes with metal ions (Tan, 1978a; Stevenson, 1976a,b). Depending on the stability of the complexes, they can be soluble or insoluble in water. Assume that the following complex reaction occurs:

$$M^{2+} + 2HA \rightleftharpoons MA_2 + 2H^+$$

where

M = metal ion
HA = humic acid
MA_2 = metal-humic acid complex

then according to the mass action law, the equilibrium constant K is

$$K = \frac{(MA_2)(H^+)^2}{(M^{2+})(HA)^2}$$

By taking the log, the equation above changes to

$$\log K = \log \frac{(MA_2)(H^+)^2}{(M^{2+})(HA)^2}$$

If the activities of HA and MA_2 are considered unity, then

$$\log K = \log \frac{(H^+)^2}{(M^{2+})}$$

or

$$\log K = 2 \log (H^+) - \log (M^{2+}) \qquad (10.1)$$

where log K is called the *stability constant*. It determines the solubility of the metal complexes (Tan et al., 1971a, Tan et al., 1971b).

Tan et al. (1971b) have calculated log K values for metal-fulvic acid complexes. They found that the stability of the complex compound is high, if the value of log K is large. Therefore, the data below indicate that Cu-FA complexes are more difficultly soluble than Zn- or Mg-FA complexes. The degree of solubility is largest for the Mg-FA complex.

Stability constants of metal-fulvic acid complexes	Cu-FA	Zn-FA	Mg-FA
Log K (pH 3.5)	7.15	5.40	3.42
Log K (pH 5.5)	8.26	5.73	4.06

Clay-Organic Compound Complexes

Clay can also form complexes with organic compounds (Tan, 1976a; Theng, 1974; 1972; Greenland, 1971). The organic compounds can be cationic, anionic, and polar nonionic in nature.

Complex Formation with Organic Anions Under ordinary conditions clay has a negative charge and, therefore, will repel organic anions. However, under certain conditions (see page 136) the broken edge surface of clay attains positive charges and will attract anions.

Negative adsorption is sometimes considered a possibility in this aspect (Theng, 1972). However, *negative adsorption* (page 172) is defined as the repulsion of anions by negatively charged clay surfaces. Therefore, the relationship between negative adsorption and complex formation (coordination bonding) is rather obscure.

A number of authors have noted the presence of negative adsorption of herbicides by montmorillonite (Bailey and White, 1970; Weber, 1970; Frissel and Bolt, 1962). Herbicides, such as 2,4-D and 2,4,5-T, are found to be negatively adsorbed by Na-montmorillonite in a medium with a pH above the pK_a value of the organic compound. Negative adsorption continues to be important, until the pH of the medium equals the pK_a, or the dissociation constant of the respective compound. Positive adsorption starts as soon as the pH is decreased below the pK_a. Bailey and White (1970) consider the dissociation constant to be a major factor in determining adsorption processes. The pK_a is used by the authors above as an indicator for the degree of acidity or basicity of the substance. At pH values $>pK_a$, these organic compounds exist in the acidic form, and behave, therefore, as anions. As such, they are subject to attraction by positive charged clay surfaces, or by negative anion adsorption. Positive charges are usually present at the edges of silicate clays, especially under strongly acid condition. Therefore, the main mechanism for adsorption of acidic organic compounds is negative adsorption. At pH values <pK_a, the compounds exist mainly in the basic form, and behave, therefore, as cations. They will be attracted by negatively charged clay surfaces.

Another possibility of interaction between acid organic compounds and silicate clays is through reactions with cations in exchange positions or through water molecules coordinated to these cations. These processes have been discussed earlier in Chapter 5.

Complex Formation with Organic Cations Under certain conditions a number of organic compounds may be positively charged, e.g., amino compounds (see pages 46 to 47). Positive charges can also develop by the following processes. Mortland (1970) suggests that after adsorption by clay, the organic compounds become positively charged by accepting protons as follows:

1. H^+-saturated clays may donate the proton

   $$R\text{-}NH_2 + H\text{-}Clay \rightleftharpoons R\text{-}NH_3\text{-}Clay$$

2. Water polarized by a cation can donate a proton to the organic compound:

   $$(M \cdot mH_2O)^{m+} + NH_2\text{-}R \rightleftharpoons (NH_3\text{-}R)^+ + [M \cdot (m - 1)H_2O]^{m-1}$$

3. By the presence of a protonated species that donates a proton to the organic molecule:

   $$\underset{\text{protonated species}}{(HA)^+} + \underset{\text{organic molecule}}{NH_2\text{-}R} \rightleftharpoons A + (NH_3\text{-}R)^+$$

These organic compounds, having attained a positive charge, may replace inorganic cations on exchange positions, or in interlayer surfaces of clays. Such an exchange follows the general laws of cation exchange reactions. The exchange occurs stoichiometricly and reaches a maximum equaling the CEC of clays (Hendricks, 1941; Greenland, 1965, 1971). If the organic cation occupies intermicellar spaces, analysis of basal spacings of clays indicates that the organic ion is adsorbed with its shortest axis perpendicular to the silicate layer (Theng, 1972).

Complex Formation with Amphoteric Organic Compounds As discussed earlier important organic substances in soils with amphoteric character are humic compounds, proteins and amino acids. The presence of functional groups, such as carboxylic and amino groups, in their molecules gives them the ability to exist either as a cation, anion or as a zwitterion (see page 46). The dominant ion species present in the soil solution depends on the soil reaction. The latter

$$CH_3-\underset{H}{\overset{NH_3^+}{C}}-COOH \rightleftarrows CH_3-\underset{H}{\overset{NH_3^+}{C}}-COO^- \rightleftarrows CH_3-\underset{H}{\overset{NH_2}{C}}-COO^-$$

Cation	Zwitterion	Anion
pH < isoelectric point	pH = isoelectric point	pH > isoelectric point

can be illustrated with the α-amino acid, L-alanine, as shown in the scheme above. The ion species in which the amino acid occurs governs the interaction with other soil components. In an acid soil reaction, or at pH values below the isoelectric point, amino acids are usually positively charged. As discussed on pages 46 to 47, the amino group can obtain an extra proton, and behave then as a cation. As such, the cationic form of amino acid can be attracted to the clay surface by cation exchange. The latter mechanism and proton transfer are expected to be the main processes for the interaction reactions at the clay-solution interface and/or in the intermicellar spaces of expanding layer silicates.

At neutral soil reactions, or soil pH close to the isoelectric point, amino acids are dipolar and behave as a zwitterion. Although the majority of soils under field conditions are in this pH range, the pH at the clay interface is generally lower than the pH of the bulk solution. Present as a zwitterion, amino acid will interact through ion dipole reactions. The positive pole (NH_3^+) can be attracted directly to the negatively charged clay surfaces. On the other hand, the negative pole of amino acid (COO^-) can also undergo interactions with metal cations adsorbed on the clay surfaces.

In alkaline conditions, or soil pH above the isoelectric point, amino acids are negatively charged and possess anion characteristics. Although the anionic form is considered to be of less importance than the cationic and zwitterionic forms (Theng, 1974), it has the capability of reacting with positively charged clay surfaces, or may be attracted to the surface of clays by cationic bridging.

Complex Formation with Nonionic Organic Compounds

The interaction between clay and uncharged organic molecules, such as alcohols and ethylene glycol, is made possible by the presence of exchangeable cations (Mortland, 1970). These cations are surrounded by water molecules arranged as a hydration shell. One of the water molecules can be exchanged for an organic ligand. The organic compound attached itself through the exchangeable cation to the clay surface. This reaction is also called *complex reaction*. If the cation has a low hydration energy, then the cation may form a direct coordinate bond with the oxygen atoms of the organic molecule.

The adsorption of organic molecules in intermicellar spaces is called intercalation. The basal spacing of clays increases stepwise as one, two, or three layers of organic compounds are intercalated. Compounds with strong polarity will orient themselves parallel to the silicate surface when intercalated (Hoffmann and Brindley, 1960).

Complex Formation and Mobility of Soil Constituents

Metal Mobility The effect of chelation on metal mobility is caused by a change in ionic behavior. After chelation the cation is surrounded by the chelating ligands. The cation may be transformed into an anion. Anions will be repelled by negatively charge colloids. Therefore, they will remain mobile. The elements, once released by weathering in the soil, behave according to the conditions. Some will be used for clay mineral synthesis, such as Si, Al, and Fe, and others will be adsorbed by soil colloids. The latter can be made available to plant growth, or be translocated, later on, with the percolating water to deeper soil depth. They may also be removed from the soil profile. For some of the elements, such as K and Na, the latter is a simple process since they form easily soluble substances. In the case of Fe and Al, it is a complex problem. Fe and Al compounds are usually insoluble at the normal soil pH range. However, the solubility of these substances can be increased, by complex formation or chelation of Fe and Al with soil humic compounds. Metals in the form of metal-chelates can percolate with the rainwater downward in the soil profile. As a soluble chelate, the metal can also be taken up by plants by exchange reaction.

Complex formation can also occur with Ca^{2+} and Mg^{2+} ions. However, since the latter are relatively soluble they can also exist as free ions. Consequently the mobility of Ca^{2+} and Mg^{2+} is less dependent on complex formation than is the case with Al^{3+} and Fe^{3+}.

Aluminum and Fe chelates are usually present in Spodosols, whereas Ca chelates occur mainly in Mollisols (Tan, 1978a). As explained earlier Al and Fe chelates have high mobility and are the reasons for formation of albic and spodic horizons. On the other hand, the Ca chelates in Mollisols are immobile. This difference in mobility is due to the difference in organic ligands. In Spodosol, the organic ligands are mostly fulvic acids, whereas in Mollisols the ligands are mainly humic acid. Fulvic acid is the water-soluble humic compound, and consequently may form soluble or easily dispersable chelates. Humic acids are insoluble in water and form stable Ca chelates in Mollisols.

Stability Diagrams of Metal Chelates

The cations in soils, such as H^+, Al^{3+}, Fe^{3+}, Zn^{2+}, Mn^{2+}, Ca^{2+}, and Mg^{2+}, may compete for bonding with the chelating agent. The cation that can form the most stable chelate in the particular condition will be bonded. Lindsay (1974) and Lindsay and Norvell (1969) have made a number of stability diagrams of Zn-, Fe-, Ca-EDTA. They reported that at pH 6-7, Zn-EDTA was stable. With the term *stable*, Lindsay and Norvell (1969) meant that the chelate existed as a soluble compound. Between pH 5.5 and pH 7.0, Fe-EDTA was the stable species in the chelate mixture (see Figure 10.1). At pH 7.0-7.3, Ca-EDTA was stable. The authors above concluded that Fe^{3+} has displaced Zn^{2+} from Zn-EDTA at low pH, while Ca^{2+} has displaced Zn^{2+} or Fe^{3+} from Zn-EDTA or Fe-EDTA at high pH. Therefore, the presence of Fe at low pH may cause Fe to become chelated and remain soluble. On the other hand, Zn is then expelled from the chelate compound and precipitates as $Zn(OH)_2$. The latter may cause Zn deficiency.

Clay Mobility

The migration of clay in soils requires that clay remains in suspension. Dispersion and suspension are increased by (1) a low electrolyte or low base content in soils, and (2) by the absence of positively charged colloids or by a combination of both factors above.

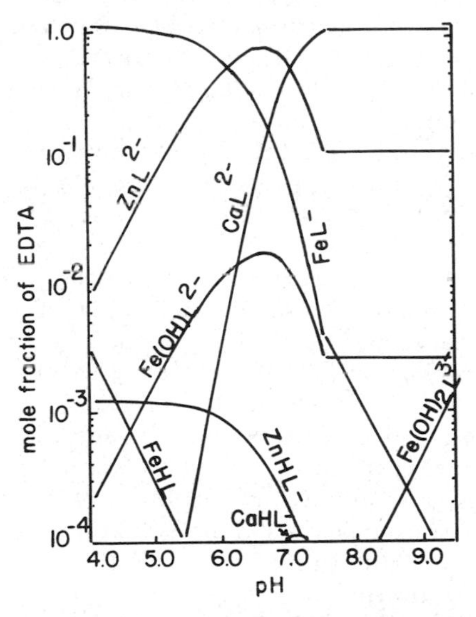

Figure 10.1 Stability diagram of Zn^{2+}, Fe^{3+}, Ca^{2+}, and H^+ complexes with EDTA and DTPA as influenced by soil pH. [Reproduced from Lindsay and Norvell (1969). By permission of the Soil Science Society of America.]

Soils high in $CaCO_3$ show little evidence of translocation of clays. However, clay can also form complexes with soil organic compounds. The latter increases its capacity to disperse (p. 207 to 208), and consequently increases its mobility.

10.3 COMPLEX FORMATION AND SOIL FERTILITY

Complex formation and chelation play an important role in improving soil feritility. In the preceding sections, it was shown that chelation increases the mobility and consequently plant availability of many cations. The release of plant nutrients by weathering of soil minerals is usually a very slow process. However, complex formation tends to accelerate the decomposition process of soil minerals and accordingly also accelerate the release of soluble nutrients.

The harmful effect of fixation of K and P is also offset by complex formation with organic ligands. Evidence has been shown by Tan (1978b) that humic and fulvic acids increase the release of K fixed in the intermicellar spaces of clays. It is expected that chelation or complex formation may also contribute towards making insoluble inorganic phosphates more soluble. The solubility of $AlPO_4$, $FePO_4$, or $Ca_3(PO_4)_2$ is expected to increase considerably by complex formation with humic compounds and/or other organic compounds. Humic and fulvic acids have a high affinity for Al, Fe, and Ca. Consequently they will compete for these elements with the respective phosphates by complex formation, releasing in the process the phosphate ions in the soil solution.

Humic compounds are also effective in binding the micronutrients, such as Fe, Cu, Zn, and Mn. In acid soils, these micronutrients are present in large amounts and cause toxicity problems to plants. By adding humus to acid soils, some of the excess of micronutrients is taken out from the solution by complex formation with humic compounds. In time they can be released again to plants in smaller amounts as needed. In this way, the chelate acts as a regulatory agent. From an environmental or ecological standpoint, complexing of heavy metal ions by humic compounds may temporarily reduce toxicity hazards for human beings, animals, and plants. In some soils, the soluble fraction of the micronutrients Fe, Zn, Cu, and Mn can be deficient since in most cases they are too insoluble. Chelation of these elements by soil organic matter increases their solubility (Lindsay, 1974). The latter helps in maintaining adequate levels of soluble micronutrients in the soil solution.

Appendix A

FUNDAMENTAL CONCEPTS

Symbol	Name	Value
c	Velocity of light	2.9979×10^{10} cm/sec
e	Electronic charge	1.6021×10^{-19} C
L	Avogadro's number	6.0225×10^{23}
h	Max Planck's constant	6.6256×10^{-27} erg sec
F	Faraday constant	96.487 C/Eq
R	Gas constant	82.056 cm^3 atm/(mol)(deg)
		1.9872 cal/(mol)(deg)
		8.3143 J/(mol)(deg)
K	Kelvin temperature	273°C = 0 K
k	Boltzmann constant	1.3805×10^{-16} erg/deg

Appendix B

GREEK ALPHABET

Greek letter	Greek name	Greek letter	Greek name
Α α	Alpha	Ν ν	Nu
Β β	Beta	Ξ ξ	Xi
Γ γ	Gamma	Ο ο	Omicron
Δ δ	Delta	Π π	Pi
Ε ε	Epsilon	Ρ ρ	Rho
Ζ ζ	Zeta	Σ σ	Sigma
Η η	Eta	Τ τ	Tau
Θ θ	Theta	Υ υ	Upsilon
Ι ι	Iota	Φ φ	Phi
Κ κ	Kappa	Χ χ	Chi
Λ λ	Lambda	Ψ ψ	Psi
Μ μ	Mu	Ω ω	Omega

Appendix C

PERIODIC CLASSIFICATION OF ELEMENTS

PERIODIC CLASSIFICATION OF ELEMENTS
based on $^{12}C = 12.0000$

Ia	2a	3b	4b	5b	6b	7b	8			Ib	2b	3a	4a	5a	6a	7a	0
1 H 1.0080																	2 He 4.003
3 Li 6.939	4 Be 9.012											5 B 10.81	6 C 12.011	7 N 14.007	8 O 15.999	9 F 18.998	10 Ne 20.183
11 Na 22.990	12 Mg 24.31											13 Al 26.98	14 Si 28.09	15 P 30.974	16 S 32.064	17 Cl 35.453	18 Ar 39.948
19 K 39.102	20 Ca 40.08	21 Sc 44.96	22 Ti 47.90	23 V 50.94	24 Cr 52.00	25 Mn 54.94	26 Fe 55.85	27 Co 58.93	28 Ni 58.71	29 Cu 63.54	30 Zn 65.37	31 Ga 69.72	32 Ge 72.59	33 As 74.92	34 Se 78.96	35 Br 79.909	36 Kr 83.80
37 Rb 85.47	38 Sr 87.62	39 Y 88.91	40 Zr 91.22	41 Nb 92.91	42 Mo 95.94	43 Tc (99)	44 Ru 101.1	45 Rh 102.90	46 Pd 106.4	47 Ag 107.87	48 Cd 112.40	49 In 114.82	50 Sn 118.69	51 Sb 121.75	52 Te 127.60	53 I 126.90	54 Xe 131.30
55 Cs 132.91	56 Ba 137.34	57 La* 138.91	72 Hf 178.49	73 Ta 180.95	74 W 183.85	75 Re 186.2	76 Os 190.2	77 Ir 192.2	78 Pt 195.09	79 Au 197.0	80 Hg 200.59	81 Tl 204.37	82 Pb 207.19	83 Bi 208.98	84 Po (210)	85 At (210)	86 Rn (222)
87 Fr (223)	88 Ra 226.05	89 Ac** (227)															

Lanthanide * series	58 Ce 140.12	59 Pr 140.91	60 Nd 144.24	61 Pm (147)	62 Sm 150.35	63 Eu 151.96	64 Gd 157.25	65 Tb 158.92	66 Dy 162.50	67 Ho 164.93	68 Er 167.26	69 Tm 168.93	70 Yb 173.04	71 Lu 174.97
Actinide ** series	90 Th 232.04	91 Pa (231)	92 U 238.03	93 Np (237)	94 Pu (242)	95 Am (243)	96 Cm (247)	97 Bk (249)	98 Cf (251)	99 Es (254)	100 Fm (253)	101 Md (256)	102 No (254)	103 Lw (257)

Appendix D

X-RAY DIFFRACTION 2θ d SPACING CONVERSION TABLE

2θ d Spacing Values for Cu Kα Radiation with λ = 1.5405 Å (0.1540 nm)

2θ	0.0	0.1	0.2	0.3	0.4	0.5	0.6	0.7	0.8	0.9
0	∞	882.63	441.32	294.21	220.66	176.53	147.11	126.09	110.33	98.076
1	88.263	80.245	73.555	67.897	63.047	58.845	55.167	51.922	49.038	46.457
2	44.135	42.033	40.122	38.378	36.779	35.308	33.950	32.693	31.526	30.440
3	29.425	28.476	27.587	26.751	25.964	25.223	24.522	23.859	23.232	22.636
4	22.071	21.532	21.020	20.531	20.065	19.619	19.193	18.785	18.394	18.018
5	17.659	17.312	16.979	16.660	16.352	16.054	15.768	15.491	15.225	14.967
6	14.717	14.476	14.243	14.017	13.798	13.586	13.381	13.181	12.988	12.800
7	12.617	12.440	12.267	12.099	11.936	11.777	11.622	11.471	11.325	11.182
8	11.042	10.906	10.773	10.644	10.517	10.394	10.273	10.155	10.040	9.9270
9	9.8168	9.7098	9.6042	9.5010	9.4001	9.3015	9.2053	9.1105	9.0173	8.9264
10	8.8378	8.7500	8.6645	8.5506	8.4989	8.4181	8.3387	8.2609	8.1847	8.1100
11	8.0360	7.9644	7.8935	7.8234	7.7549	7.6880	7.6220	7.5571	7.4932	7.4305
12	7.3688	7.3081	7.2484	7.1897	7.1320	7.0751	7.0192	6.9642	6.9100	6.8567
13	6.8042	6.7524	6.7015	6.6513	6.6019	6.5532	6.5053	6.4550	6.4114	6.3655
14	6.3203	6.2757	6.2317	6.1883	6.1456	6.1035	6.0619	6.0209	5.9804	5.9405
15	5.9011	5.8623	5.8239	5.7860	5.7488	5.7119	5.6755	5.6395	5.6041	5.5691
16	5.5345	5.5004	5.4666	5.4333	5.4004	5.3679	5.3358	5.3040	5.2727	5.2417
17	5.2111	5.1809	5.1510	5.1214	5.0922	5.0633	5.0348	5.0065	4.9787	4.9511
18	4.9238	4.8968	4.8701	4.8437	4.8176	4.7918	4.7663	4.7410	4.7160	4.6913
19	4.6669	4.6426	4.6187	4.5950	4.5715	4.5482	4.5253	4.5026	4.4801	4.4577
20	4.4357	4.4138	4.3922	4.3708	4.3496	4.3287	4.3079	4.2872	4.2669	4.2467
21	4.2267	4.2069	4.1872	4.1678	4.1486	4.1295	4.1106	4.0919	4.0733	4.0550
22	4.0367	4.0187	4.0008	3.9831	3.9656	3.9481	3.9309	3.9139	3.8969	3.8801
23	3.8635	3.8469	3.8306	3.8144	3.7983	3.7824	3.7666	3.7509	3.7354	3.7200
24	3.7047	3.6896	3.6746	3.6596	3.6449	3.6302	3.6157	3.6013	3.5870	3.5728
25	3.5587	3.5448	3.5309	3.5172	3.5036	3.4901	3.4767	3.4634	3.4502	3.4371
26	3.4241	3.4112	3.3984	3.3857	3.3731	3.3606	3.3482	3.3359	3.3236	3.3115
27	3.2995	3.2875	3.2758	3.2639	3.2522	3.2406	3.2291	3.2176	3.2063	3.1951
28	3.1839	3.1727	3.1617	3.1508	3.1399	3.1291	3.1184	3.1078	3.0973	3.0868
29	3.0763	3.0660	3.0557	3.0455	3.0354	3.0253	3.0153	3.0054	2.9955	2.9857
30	2.9760	2.9664	2.9568	2.9472	2.9377	2.9283	2.9190	2.9098	2.9005	2.8914
31	2.8823	2.8732	2.8643	2.8553	2.8465	2.8376	2.8289	2.8202	2.8116	2.8029
32	2.7945	2.7859	2.7775	2.7691	2.7608	2.7526	2.7443	2.7362	2.7281	2.7200
33	2.7120	2.7040	2.6961	2.6882	2.6804	2.6727	2.6649	2.6573	2.6496	2.6420
34	2.6345	2.6270	2.6195	2.6121	2.6048	2.5974	2.5902	2.5830	2.5757	2.5686
35	2.5615	2.5541	2.5474	2.5404	2.5334	2.5295	2.5196	2.5129	2.5060	2.4993
36	2.4926	2.4859	2.4793	2.4727	2.4661	2.4596	2.4531	2.4466	2.4402	2.4338
37	2.4274	2.4211	2.4149	2.4086	2.4024	2.3962	2.3901	2.3840	2.3779	2.3719
38	2.3659	2.3599	2.3540	2.3480	2.3421	2.3362	2.3305	2.3247	2.3189	2.3131
39	2.3074	2.3018	2.2962	2.2905	2.2849	2.2794	2.2739	2.2684	2.2629	2.2574

Appendix E

SYSTEM INTERNATIONAL (SI) UNITS

SI unit	Symbol
Ampere (electrical current)	A
Candela (luminous intensity)	cd
Meter (length)	m
Mole (amount of substance)	mol
Kelvin (thermodynamic temperature)	K
Kilogram (mass)	kg
Second (time)	s
Square meter (area)	m^2

Factors for Converting into SI Units

U.S. unit	SI unit	To obtain SI unit multiply U.S. unit by
Acre	Hectare, ha	0.405
Acre	Square meter, m^2	4.05×10^3
Atmosphere	Megapascal, MPa	0.101
Calorie	Joule, J	4.19
Cubic foot	Liter, L	28.3

Factors for Converting into SI Units (Continued)

U.S. unit	SI unit	To obtain SI unit multiply U.S. unit by
Cubic inch	Cubic meter, m^3	1.64×10^{-5}
Curie	Becquerel, Bq	3.7×10^{10}
Dyne	Newton, N	10^{-5}
Erg	Joule, J	10^{-7}
Foot	Meter, m	0.305
Gallon	Liter, L	3.78
Gallon per acre	Liter per ha	9.35
Inch	Centimeter, cm	2.54
Mile	Kilometer, km	1.61
Miles per hour	Meter per second	0.477
Ounce (weight)	Gram, g	28.4
Ounce (fluid)	Liter, L	2.96×10^{-2}
Pint	Liter, L	0.473
Pound	Gram, g	454
Pound per acre	Kilogram per ha	1.12
Pound per cubic foot	Kilogram per m^3	16.02
Pound per square foot	Pascal, Pa	47.9
Pound per square inch	Pascal, Pa	6.9×10^3
Quart	Liter, L	0.946
Square foot	Square meter, m^2	9.29×10^{-2}
Square inch	Square cm, cm^2	6.45
Square mile	Square kilometer, km^2	2.59
Ton	Kilogram, kg	907
Ton per acre	Megagram per ha	2.24

REFERENCES AND ADDITIONAL READINGS

Adler, H. H., P. F. Kerr, E. E. Bray, N. P. Stevens, J. M. Hunt, W. D. Keller, and E. E. Pickett (1950). Infrared spectra of reference clay minerals. Amer. Petroleum Inst., Project 49. Clay Mineral Standards Prelim. Report No. 8. Cornell University, Ithaca, N.Y.

Aslyng, H. C. (1963). Soil physics terminology. Inter. Soc. Soil Sci. Bull. *23*:7-10.

Bailey, G. W., and J. L. White (1970). Factors influencing the adsorption and movement of pesticides in soils. In *Residue Reviews*, F. A. Gunther and J. D. Gunther (eds.), Vol 32, Springer-Verlag, New York, pp 29-92.

Barnhisel, R. I. (1977). Chlorites and hydroxy interlayered vermiculite and smectite. In *Minerals in Soil Environments*, J. B. Dixon, S. B. Weed, J. A. Kittrick, M. H. Milford, and J. L. White (eds.), Soil Sci. Soc. Amer., Madison, Wis., pp. 331-356.

Baver, L. D. (1963). The effect of organic matter on soil structure. Pontificiae Academiae Scientiarvm Scripta Varia *32*:383-413.

Bernstein, L., and L. E. Francois (1973). Leaching requirement studies: Sensitivity of alfalfa to salinity of irrigation and drainage waters. Soil Sci. Soc. Amer. Proc. *37*:931-943.

Besoain, Eduard (1968). Imogolite in volcanic soils of Chile. Geoderma *2*:151-169.

Birkeland, P. W. (1974). *Pedology, Weathering, and Geomorphological Research*. Oxford University Press, New York.

Bjerrum, Niels (1923). Die Konstitution der Ampholyte, besonders der Aminosäuren, und ihre Dissoziations-Konstanten. Z. Physikalische Chemie *104*:147-173.

Bloomfield, C. (1951). Experiments on the mechanism of gley formation. J. Soil Sci. *2*:196-211.

Bloomfield, C. (1953). A study of podzolization. Part 1. The mobilization of iron and aluminum by Scots pine needles. J. Soil Sci. *4*:5-16.

Bolt, G. H. (1967). Cation-exchange equations used in soil science--A review. Neth. J. Agric. Sci. *15*:81-103.

Bolt, G. H. (1976). Adsorption of anions by soils. In *Soil Chemistry. A. Basic Elements*, G. H. Bolt and M. G. M. Bruggenwert (eds.), Elsevier Sci., Amsterdam, pp. 91-95.

Bolt, G. H., M. G. M. Bruggenwert, and A. Kamphorst (1976). Adsorption of cations by soil. In *Soil Chemistry. A. Basic Elements*, G. H. Bolt and M. G. M. Bruggenwert (eds.), Elsevier Sci., Amsterdam, pp. 54-95.

Brady, Nyle C. (1974). *The Nature and Properties of Soils*. 8th ed., Macmillan New York.

Bragg, Sir L., and G. F. Claringbull (1965). *Crystal Structures of Minerals*. Bell, London.

Brindley, G. W., S. W. Bailey, G. T. Faust, S. A. Forman, and C. I. Rich (1968). Report of the nomenclature committee (1966-1967) of the Clay Minerals Society. Clays and Clay Minerals *16*:322-324.

Brunauer, S., P. H. Emmett, and E. Teller (1938). Adsorption of gases in multimolecular layers. J. Amer. Chem. Soc. *60*:309.

Buol, S. W., R. J. McCracken, and F. D. Hole (1973). *Soil Genesis and Classification*. The Iowa State University Press, Ames, Iowa.

Burges, A. (1960). The nature and distribution of humic acid. Scient. Proc. Royal Dublin Soc. Series A *1*:53-59.

Chapman, D. L. (1913). A contribution to the theory of electrocapillarity. Phil. Mag. *25(6)*:475-481.

Chen, Y., and M. Schnitzer (1976). Scanning electron microscopy of a humic acid and its metal and clay complexes. Soil Sci. Soc. Amer. J. *40*:682-686.

Chen, Y., N. Senesi, and M. Schnitzer (1977). Information provided on humic substances by E_4/E_6 ratios. Soil Sci. Soc. Amer. J. *41*:352-358.

Chen, Y., N. Senesi, and M. Schnitzer (1978). Chemical and physical characteristics of humic and fulvic acids extracted from soils of the Mediterranean region. Geoderma *20*:87-104.

Clark, F. E., and K. H. Tan (1969). Identification of a polysaccharide ester linkage in humic acid. Soil Biol. Biochem. *1*:75-81.

Cranwell, P. A., and R. D. Haworth (1975). The chemical nature of humic acids. In *Humic Substances, Their Structure and Function in the Biosphere*, D. Povoledo and H. L. Golterman (eds.), Centre for Agric. Publ. and Documentation, Wageningen, pp. 13-18.

Davies, C. W. (1962). *Ion Association*. Butterworth, Washington, D.C.

Davis, L. E. (1945). Simple kinetic theory of ionic exchange for ions of unequal charge. J. Phys. Chem. *49*:473-479.

Dixon, J. B. (1977). Kaolinite and serpentine group minerals. In *Minerals in Soil Environment*, J. B. Dixon, S. B. Weed, J. A. Kittrick, M. H. Milford, and J. L. White (eds.). Soil Sci. Soc. Amer. Madison, Wisconsin, pp. 357-403.

Donahue, Roy L., Raymond W. Miller, and John C. Shickluna (1977). *Soils, An Introduction to Soils and Plant Growth.* Prentice-Hall, Englewood Cliffs, N.J.

Dormaar, J. F. (1974). Scanning electron microscopy as applied to organo-mineral complexes in alkaline extracts of soil. Soil Sci. Soc. Amer. Proc. *38*:685-686.

Dormaar, J. F. (1975). Effects of humic substances from Chernozemic A_h horizons on nutrient uptake by *Phaseolus vulgaris* and *Festuca sabrella.* Can. J. Soil Sci. *55*:111-118.

Douglas, L. A. (1977). Vermiculites. In *Minerals in Soil Environments*, J. B. Dixon, S. B. Weed, J. A. Kittrick, M. H. Milford, and J. L. White (eds.). Soil Sci. Soc. Amer., Madison, Wisconsin, pp. 259-292.

Egawa, T., and Y. Watanabe (1964). Electron micrographs of the clay minerals in Japanese soils. Bull. Nat. Int. Agri. Sci. (Japan), Series B *14*.

Eltantawy, I. M. (1980). The effect of heating on humic acid structure and electron spin resonance signal. Soil Sci. Soc. Amer. J. *44*:512-514.

Eltantawy, I. M., and M. Baverez (1978). Structural study of humic acids by x-ray, electron spin resonance, and infrared spectroscopy. Soil Sci. Soc. Amer. J. *42*:903-905.

Eriksson, E. (1952). Cation exchange equilibria on clay minerals. Soil Sci. *74*:103-113.

Eswaran, H. (1972). Morphology of allophane, imogolite and a halloysite. Clay Minerals *9*:281-285.

Eswaran, H., G. Stoops, and C. Sys (1977). The micromorphology of gibbsite forms in soils. J. Soil Sci. *28*:136-143.

Evans, R. C. (1939). *An Introduction to Crystal Chemistry.* Cambridge University Press, London.

Fanning, D. S., and V. Z. Keramidas (1977). Micas. In *Minerals in Soil Environments*, J. B. Dixon, S. B. Weed, J. A. Kittrick, M. H. Milford, and J. L. White (eds.). Soil Sci. Soc. Amer., Madison, Wisconsin, pp. 195-258.

Farmer, V. C. (1968). Infrared spectroscopy in clay mineral studies. Clay Minerals *7*:373-387.

Farmer, V. C., and F. Palmieri (1975). The characterization of soil minerals by infrared spectroscopy. In *Soil Components. Vol. 2. Inorganic Components*, J. E. Gieseking (ed.). Springer-Verlag, New York, pp. 573-671.

Felbeck Jr., G. T. (1965). Structural chemistry of soil humic substances. Advan. Agronomy *17*:327-368.

Flaig, W. (1975). An introductory review on humic substances: aspects of research on their genesis, their physical and chemical properties, and their effect on organisms. In *Humic Substances, Their Structure and Function in the Biosphere*, D. Povoledo and

H. L. Golterman (eds.). Centre for Agric. Publ. and Documentation, Wageningen, pp. 19-42.

Flaig, W., and H. Beutelspacher (1951). Electron microscope investigations on natural and synthetic humic acids. Z. Pflanzen Ernährung, D. B. *52*:1-21.

Flaig, W., H. Beutelspacher, and E. Rietz (1975). Chemical composition and physical properties of humic substances. In *Soil Components. Vol. 1. Organic Components*, J. E. Gieseking (ed.). Springer-Verlag, New York, pp. 1-211.

Foster, M. D. (1962). Interpretation of the composition and a classification of the chlorites. U.S. Geol. Survey Professional Paper 414A.

Foth, H. D., and L. M. Turk (1978). *Fundamentals of Soil Science.* Wiley, New York

Freundlich, H. (1926). *Colloid and Capillary Chemistry.* Methuen, London.

Frissel, M. J., and G. H. Bolt (1962). Interaction between certain ionizable organic compounds (herbicides) and clay minerals. Soil Sci. *94*:284-291.

Gapon, E. N. (1933). Theory of exchange adsorption in soils. J. Gen. Chem. (USSR) *3*(2):144-152.

Garrels, R. M., and C. L. Christ (1965). *Solutions, Minerals, and Equilibria.* Harper and Row, New York.

Gast, R. C. (1977). Surface and colloid chemistry. In *Minerals in Soils Environment*, J. B. Dixon, S. B. Weed, J. A. Kittrick, M. H. Milford, and J. L. White (eds.). Soil Sci. Soc. Amer., Madison, Wis., pp. 27-73.

Glasstone, S. (1946). *Textbook of Physical Chemistry.* Van Norstrand, Princeton, N.J.

Goh, K. M., and F. J. Stevenson (1971). Comparison of infrared spectra of synthetic and natural humic and fulvic acids. Soil Sci. *112*:392-400.

Gosh, K., and M. Schnitzer (1980a). Effects of pH and neutral electrolyte concentration on free radicals in humic substances. Soil Sci. Soc. Amer. J. *44*:975-978.

Gosh, K., and M. Schnitzer (1980b). Fluorescence excitation spectra of humic substances. Can. J. Soil Sci. *60*:373-379.

Gosh, K., and M. Schnitzer (1980c). Macromolecular structures of humic substances. Soil Sci. *129*:266-276.

Gouy, G. (1910). Sur la constitution de la charge électrique à la surface d'un électrolyte. Ann. Phys. (Paris), Series 4, *9*:457-468.

Greenland, D. J., G. R. Lindstrom, and J. P. Quirk (1961). Role of polysaccharides in stabilisation of natural soil aggregates. Nature (London) *191*:1283-1284.

Greenland, D. J., G. R. Lindstrom, and J. P. Quirk (1962). Organic materials which stabilize natural soil aggregates. Soil Sci. Soc. Amer. Proc. *26*:366-371.

Greenland, D. J., and J. P. Quirk (1962). Surface areas of soil colloids. Trans. Intern. Soil Sci. Conf., Comm. IV and V, Palmerston North, New Zealand, pp. 79-87.

Greenland, D. J., and J. P. Quirk (1964). Determination of the total surface areas of soils by adsorption of cetyl pyridium bromide. J. Soil Sci. *15*:178-191.

Greenland, D. J. (1965). Interaction between clays and organic compounds in soils. 1. Mechanisms of interaction between clays and defined organic compounds. Soils and Fert. *28*:415-425.

Greenland, D. J. (1971). Interactions between humic and fulvic acids and clays. Soil Sci. *111*:34-41.

Greenland, D. J., and M. H. B. Hayes (eds.) (1978). *The Chemistry of Soil Constituents.* Wiley-Interscience, New York.

Grimshaw, R. W. (1971). *The Chemistry and Physics of Clays.* Wiley-Interscience, New York.

Guminski, S. (1957). The mechanism and conditions of the physiological actions of humic substances on the plant. Pochvovedenie *12*:36.

Guminski, S., and Z. Guminska (1953). Studies on the activity of humus on plants. Acta Societatis Botanicorum Poloniae *22*:45-54.

Guminski, S., D. Augustin, and J. Sulej (1977). Comparison of some chemical and physico-chemical properties of natural and model sodium humates and of biological activity of both substances in tomato water cultures. Acta Societatis Botanicorum Poloniae *XIVI*:437-448.

Hem, J. D., and W. H. Cropper (1959). Survey of ferrous-ferric chemical equilibria and redox potentials. U.S. Geol. Survey Water Supply Paper 1459A.

Hendricks, S. B. (1941). Base exchange of the clay mineral montmorillonite for organic cations and its dependence upon adsorption due to Van der Waals forces. J. Phys. Chem. *45*:65-81.

Hillel, D. (1972). *Soil and Water. Physical Principles and Processes.* Academic Press, New York.

Hodgson, J. F. (1969). Metal-organic complexing agents and transport of metal to roots. Soil Sci. Soc. Amer. Proc. *33*:68-75.

Hoffmann, R. W., and G. W. Brindley (1960). Adsorption of non-ionic aliphatic molecules from aqueous solutions on montmorillonite. Clay-organic studies II. Geochim. Cosmochim. Acta *20*:15-29.

Holty, J. G., and P. E. Heilman (1971). Molecular sieve fractionation of organic matter in a Podzol from southeastern Alaska. Soil Sci. *112*:351-356.

Inoue, Takahiro, and Koji Wada (1973). Adsorption of humified clover extracts by various clays. *Trans. 9th Inter. Congress Soil Sci., Adelaide, Australia, 1968. Vol. III*, American Elsevier, New York, pp. 289-298.

Jackson, M. L., and G. D. Sherman (1953). Chemical weathering of minerals in soils. Adv. Agronomy 5:219-318.

Jeffery, J. W. O. (1960). Iron and the E_h of waterlogged soils with particular reference to paddy. J. Soil Sci. *11*:140-148.

Jenny, H. (1936). Simple kinetic theory of ionic exchange. I. Ions of equal valency. J. Phys. Chem. *40*:501-517.

Kamphorst, A., and G. H. Bolt (1976). Saline and sodic soils. In *Soil Chemistry. A. Basic Elements*, G. H. Bolt and M. G. M. Bruggenwert (eds.). Elsevier, Amsterdam, pp. 171-191.

Keller, W. D. (1954). Bonding energies of some silicate minerals. Amer. Mineralogist *39*:783-793.

Keller, W. D. (1964). Processes of origin and alteration of clay minerals. In *Soil Clay Mineralogy. A Symposium*, C. I. Rich and G. W. Kunze, (eds.). The University of North Carolina Press, Chapel Hill, N.C., pp. 3-76.

Kittrick, J. A. (1967). Gibbsite-kaolinite equilibria. Soil Sci. Soc. Amer. Proc. *31*:314-316.

Kittrick, J. A. (1969). Soil minerals in the $Al_2O_3—SiO_2—H_2O$ system and a theory of their formation. Clays and Clay Minerals *17*:157-167.

Klotz, I. M. (1950). *Chemical Thermodynamics*. Prentice-Hall, Englewood Cliffs, N.J.

Kolthoff, I. M., and E. B. Sandell (1952). *Textbook of Quantitative Inorganic Analysis*. Macmillan, New York.

Kononova, M. M. (1961). *Soil Organic Matter*, T. Z. Nowakowski and G. A. Greenwood (trans.). Pergamon, Oxford.

Kononova, M. M. (1966). *Soil Organic Matter*. Pergamon, Oxford, pp. 400-404.

Krauskopf, K. B. (1956). Dissolution and precipitation of silica at low temperatures. Geochim. Cosmochim. Acta *10*:1-26.

Krishnamoorty, C., and R. Overstreet (1949). Theory of ion exchange relationships. Soil Sci. *68*:307-315.

Kruyt, H. R. (1944). *Inleiding tot de Physische Chemie*. Uitgeverij H. J. Paris, Amsterdam.

Kumada, K. (1965). Studies on the colour of humic acids. Part 1. On the concepts of humic substances and humification. Soil Sci. Plant Nutr. *11*:151-156.

Kumada, K., and H. M. Hurst (1967). Green humic acid and its possible origin as a fungal metabolite. Nature *214*:5-88.

Kumada, K., and E. Miyara (1973). Sephadex gel filtration of humic acids. Soil Sci. Plant Nutr. *19*:255-263.

Lagerwerff, J. V., and G. H. Bolt (1959). Theoretical and experimental analysis of Gapon's equation for ion exchange. Soil Sci. *87*:127-222.

Langmuir, Irving (1918). The adsorption of gases on plane surfaces of glass, mica and platinum. J. Amer. Chem. Soc. *40*:1361-1382.

Lewis, G. N., and M. Randall (1921). The activity coefficient of strong electrolytes. J. Amer. Chem. Soc. *43*:1112.

Lindsay, W. L. (1974). Role of chelation in micronutrient availability. In *The Plant Root and Its Environment*, E. W. Carson (ed.). University Press of Virginia, Charlottesville, Va., pp. 507-524.

Lindsay, W. L., and W. A. Norvell (1969). Equilibrium relationships of Zn^{2+}, Fe^{3+}, Ca^{2+}, and H^+ with EDTA and DTPA in soils. Soil Sci. Soc. Amer. Proc. *33*:62-68.

Lindsay, W. L., M. Peech, and J. S. Clark (1959). Solubility criteria for the existence of variscite in soils. Soil Sci. Soc. Amer. Proc. *23*:357-360.

Lowe, L. E., and W-C. Tsang (1970). Distribution of a green humic acid component in forest humus layers of British Columbia. Can. J. Soil Sci. *50*:456-457.

MacEwan, D. M. C., and A. Ruiz-Amil (1975). Interstratified clay minerals. In *Soil Components, Vol. 2, Inorganic Components*, J. E. Gieseking (ed.). Springer-Verlag, New York, pp. 265-334.

Mackenzie, R. C. (1975). The classification of soil silicates and oxides. In *Soil Components, Vol. 2, Inorganic Components*, J. E. Gieseking (ed.). Springer-Verlag, New York, pp. 1-25.

Mackenzie, R. C., and S. Caillere (1975). The thermal characteristics of soil minerals and the use of these characteristics in the qualitative and quantitative determination of clay minerals in soils. In *Soil Components, Vol. 2, Inorganic Components*, J. E. Gieseking (ed.). Springer-Verlag, New York, pp. 529-571.

Manov, G. G., R. G. Bates, W. J. Hamer, and S. F. Acree (1943). Values of the constants in the Debye-Hückel equation for activity coefficients. J. Am. Chem. Soc. *65*:1765-1767.

Martell, A. E., and M. Calvin (1952). *Chemistry of the Metal Chelate Compounds*. Prentice-Hall, New York

Martin, A. E., and R. Reeve (1960a). Chemical studies of podzolized illuvial horizons. V. The flocculation of humus by ferric and ferrous iron and by nickel. J. Soil Sci. *11*:382-393.

Martin, A. E., and R. Reeve (1960b). Chemical studies of podzolic illuvial horizons. IV. The flocculation of humus by aluminum. J. Soil Sci. *11*:369-381.

Martin, J. P., J. O. Ervin, and R. A. Shepherd (1966). Decomposition of the iron, aluminum, zinc and copper salts or complexes of some microbial and plant polysaccharides in soil. Soil Sci. Soc. Amer. Proc. *30*:196-200.

Martin, J. P., K. Haider, and E. Bondietto (1975). Properties of model humic acids synthesized by phenoloxidase and autoxidation of phenols and other compounds formed by soil fungi. In *Humic Substances, Their Structure and Function in the Biosphere*, D. Povoledo and H. Golterman (eds.). Centre for Agric. Publ. and Documentation, Wageningen, The Netherlands, pp. 171-186.

Martin, F., C. Saiz-Jimenes, and A. Cert (1977). Pyrolysis-gas chromatography-mass spectrometry of soil humic fractions. I. The low boiling point compounds. Soil Sci. Soc. Amer. J. *41*:1114-1118.

Mehta, N. C., P. Dubach, and H. Deuel (1963). Untersuchungen Uber die Molekular gewichts Verteilung von Huminstoffen durch Gelfiltration an Sephadex. Z. Pflanzenern. Dung. B. *102*:128-137.

Mekaru, T., and G. Uehara (1972). Anion adsorption in ferruginous tropical soils. Soil Sci. Soc. Amer. Proc. *36*:296-300.

Mellor, D. P. (1964). Historical background and fundamental concepts. In *Chelating Agents and Metal Chelates*, F. P. Dwyer and D. P. Mellor (eds.). Academic Press, New York, pp. 1-50.

Millot, G. (1970). *Geology of Clays*. Springer-Verlag, New York.

Mortenson, J. L. (1961). Physico-chemical properties of a soil polysaccharide. *Trans. 7th Int. Congress Soil Sci., 1960 Madison, Wisconsin.* American Elsevier, New York, II:98-104.

Mortenson, J. L., D. M. Anderson, and J. L. White (1965). Infrared spectrometry. In *Methods of Soil Analysis*, Part 1, C. A. Black, D. D. Evans, J. L. White, L. E. Ensminger, and F. E. Clark (eds.), Agronomy Series No. 9. Amer. Soc. Agronomy, Madison, Wisconsin, pp. 743-770.

Mortland, M. M. (1970). Clay-organic complexes and interactions. Advances Agronomy. *22*:75-117.

Murmann, R. K. (1964). *Inorganic Complex Compounds*. Reinhold, New York.

Norrish, K. (1954). Swelling of montmorillonite. Disc. Faraday Soc. *18*:120-134.

Novozamsky, I., J. Beek, and G. H. Bolt (1976). Chemical Equilibria. In *Soil Chemistry. A. Basic Elements*, G. H. Bolt and M. G. M. Bruggenwert (eds.). Elsevier Scient., Amsterdam, pp. 13-42.

Ogner, G. (1979). The ^{13}C nuclear magnetic resonance spectrum of a methylated humic acid. Soil Biol. Biochem. *11*:105-108.

Ogner, G. (1980). Analysis of the carbohydrates of fulvic and humic acids as their partially methylated alditol acetates. Geoderma *23*: 1-10.

Olness, A., and C. E. Clapp (1973). Occurrence of collapsed and expanded crystals in montmorillonite-dextran complexes. Clays and Clay Minerals *21*:289-293.

Olness, A., and C. E. Clapp (1975). Influence of a polysaccharide structure on dextran adsorption by montmorillonite. Soil Biol. Biochem. 7:113-118.

Orioli, G. A., and N. R. Curvetto (1980). Evaluation of extractants for soil humic substances. 1. Isotachophoretic studies. Plant Soil *55*:353-361.

Orlov, D. S., and N. L. Erosiceva (1967). *Zur frage der Wechselwirkung von Huminsauren mit einigen Metallkationen.* Vestnik-Moskovskogo Universiteta. Biologia, Pochvovedinie 1, pp. 98-105.

Orlov, D. S., Ya. M. Ammosova, and G. I. Glebova (1975). Molecular parameters of humic acids. Geoderma *13*:211-229.

Paton, T. R. (1978). *The Formation of Soil Material.* Allen & Unwin, Boston, Mass.

Pauling, L. (1929). The principles underlying the structure of complex ionic compounds. J. Amer. Chem. Soc. *51*:1010-1026.

Perrin, D. D. (1964). *Organic Complexing Reagents.* Interscience, New York.

Pierce, R. H., Jr., and G. T. Felbeck, Jr. (1975). A comparison of three methods of extracting organic matter from soils and marine sediments. In *Humic Substances, Their Structure and Function in the Biosphere,* D. Povoledo and H. L. Golterman (eds.). Centre for Agric. Publ. and Documentation, Wageningen, The Netherlands, pp. 217-232.

Poapst, P. A., C. Genier, and M. Schnitzer (1970). Effect of a soil fulvic acid on stem elongation in peas. Plant and Soil *32*:367-372.

Poapst, P. A., C. Genier, and M. Schnitzer (1971). Fulvic acid and adventitious root formation. Soil Biol. Biochem. *3*:367-372.

Posner, A. M. (1964). Titration curves of humic acid. Trans. Intern. Congress Soil Science, Bucarest, Romania, 1964. Academy of the Socialist Republic of Romania, Bucharest, *11*:161-164.

Rich, C. I. (1968). Mineralogy of soil potassium. In *The Role of Potassium in Agriculture,* V. J. Kilmer, S. E. Younts, and N. C. Brady (eds.). Amer. Soc. Agron. Madison, Wis., pp. 79-108.

Richards, L. A. (ed.) (1954). *Diagnosis and Improvement of Saline and Alkali Soils.* USDA Agriculture Handbook No. 60. U.S. Government Printing Office, Washington, D.C.

Riffaldi, R., and M. Schnitzer (1972). Electron spin resonance spectrometry of humic substances. Soil Sci. Soc. Amer. Proc. *36*:301-307.

Ross, C. S., and P. F. Kerr (1934). Halloysite and allophane. U.S. Geol. Survey Professional Paper 185G, pp. 135-148.

Ruggiero, P., F. S. Interesse, and O. Sciacovelli (1980). ^{1}H NMR evidence of exchangeable aromatic protons in fulvic and humic acids. Soil Biol. Biochem. *12*(3):297-299.

Salfeld, J. Chr. (1975). Ultraviolet and visible adsorption spectra of humic systems. In *Humic Substances, Their Structure and Function in the Biosphere,* D. Povoledo and H. L. Golterman (eds.). Centre for Agric. Publ. and Documentation, Wageningen, The Netherlands, pp. 269-280.

Schnitzer, M. (1965). The application of infrared spectroscopy to investigations on soil humic compounds. Can. Spectrosc. *10*:121-127.

Schnitzer, M. (1971). Characterization of humic constituents by spectroscopy. In *Soil Biochemistry, Vol. 2,* A. D. McLaren and J. Skujins (eds.). Dekker, New York, pp. 60-95.

Schnitzer, M. (1974). The methylation of humic substances. Soil Sci. *117*:94-102.

Schnitzer, M. (1975). Chemical, spectroscopic and thermal methods for the classification and characterization of humic substances. In *Humic Substances, Their Structure and Function in the Biosphere*, D. Povoledo and H. L. Golterman (eds.). Centre for Agric. Publ. and Documentation, Wageningen, The Netherlands, pp. 293-310.

Schnitzer, M. (1976). The Chemistry of Humic Substances. In *Environmental Biogeochemistry. Vol. 1. Carbon, Nitrogen, Phosphorus, Sulfur, and Selenium Cycles*, J. O. Nriagu (ed.), Proc. 2d Intern. Symposium on Environmental Biogeochemistry, Hamilton, Ontario, Canada, April 8-12, 1975. Ann Arbor Sci., Ann Arbor, Mich., pp. 89-107.

Schnitzer, M., and D. A. Hindle (1980). Effect of peracetic acid oxidation on N-containing components of humic materials. Can. J. Soil Sci. *60*:541-548.

Schnitzer, M., and S. U. Khan (1972). *Humic Substances in the Environment*. Dekker, New York.

Schnitzer, M., and H. Kodama (1976). The dissolution of micas by fulvic acid. Geoderma *15*:381-391.

Schnitzer, M., D. A. Shearer, and J. R. Wright (1959). A study in the infrared of high-molecular-weight organic matter extracted by various reagents from a Podzolic B horizon. Soil Sci. *87*:252-257.

Schofield, R. K. (1947). A ratio law governing the equilibrium of cations in the soil solution. Proc. 11th Int. Congr. Pure Appl. Chem. *3*:257-261.

Schofield, R. K. (1955). Can a precise meaning be given to "available phosphorus"? Soils and Fert. *18*:373-375.

Schofield, R. K., and A. W. Taylor (1955). The measurement of soil pH. Soil Sci. Soc. Amer. Proc. *19*:164-167.

Schwertmann, U., and R. M. Taylor (1977). Iron oxides. In *Minerals in Soil Environments*, J. B. Dixon, S. B. Weed, J. A. Kittrick, M. H. Milford, and J. L. White (eds.), Soil Sci. Soc. Amer. Madison, Wis., pp. 145-180.

Senesi, N., and M. Schnitzer (1977). Effects of pH, reaction time, chemical reduction and irradiation on ESR spectra of fulvic acid. Soil Sci. *123*:224-234.

Simonson, R. W. (1959). Outline of a generalized theory of soil genesis. Soil Sci. Soc. Amer. Proc. *23*:152-156.

Slatyer, R. O. (1957). The significance of the permanent wilting percentage in studies of plant and soil water relations. Bot. Rev. *23*:585-636.

Slatyer, R. O. (1967). *Plant-water Relationships*. Academic Press, New York.

Sposito, G., and K. M. Holtzclaw (1977). Titration studies on the polynuclear, polyacidic nature of fulvic acid extracted from sewage sludge-soil mixtures. Soil Sci. Soc. Amer. J. *41*:330-336.

Steelink, C. (1964). Free radical studies of lignin, lignin degradation products and soil humic acids. Geochim. Cosmochim. Acta *28*: 1615-1622.

Steelink, C., and G. Tollin (1967). Free radicals in soils. In *Soil Biochemistry*, A. D. McLaren and G. H. Peterson (eds.). Marcel Dekker Inc. New York, Chapter 6, pp. 147-173.

Stern, O. (1924). Zür Theory der elektrolytischen Doppelschicht. Z. Elektrochem. *30*:508-516.

Stevenson, F. J. (1965). Gross chemical fractionation of organic matter. In *Methods of Soil Analysis*, Part 2, C. A. Black, D. D. Evans, J. L. White, L. E. Ensminger, and F. E. Clark (eds.), Agronomy Series No. 9. Amer. Soc. Agronomy, Madison, Wis., pp. 1409-1421.

Stevenson, F. J. (1976a). Stability constants of Cu^{2+}, $Pb^{2+}$and Cd^{2+} complexes with humic acids. Soil Sci. Soc. Amer. J. *40*:665-672.

Stevenson, F. J. (1976b). Binding of metal ions by humic acids. In *Environmental Biogeochemistry. Vol. 1. Carbon, Nitrogen, Phosphorus, Sulfur, and Selenium Cycles*. J. O. Nriagu (ed.). Proc. 2d. Intern. Symposium on Environmental Biogeochemistry, Hamilton, Ontario, Canada, April 8-12, 1975. Ann Arbor Sci., Ann Arbor, Mich., pp. 519-540.

Stevenson, F. J., and K. M. Goh (1971). Infrared spectra of humic acids and related substances. Geochim. Cosmochim. Acta *35*:471-483.

Stevenson, F. J., and K. M. Goh (1972). Infrared spectra of humic and fulvic acid and their methylated derivatives, evidence for nonspecificity of analytical methods for oxygen-containing functional groups. Soil Sci. *113*:334-345.

Sticher, H., and R. Bach (1966). Fundamentals in the chemical weathering of silicates. Soils and Fert. *29*:321-325.

Sudo, T., and H. Yotsumoto (1977). The formation of halloysite tubes from spherulitic halloysite. Clays and Clay Min. *25*:155-159.

Swift, R. S., B. K. Thornton, and A. M. Posner (1970). Spectral characteristics of a humic acid fractionated with respect to molecular weight using agar gel. Soil Sci. *110*:93-99.

Swindale, L. D. (1975). The crystallography of minerals of the kaolin group. In *Soil Components, Vol. 2, Inorganic Components*, J. E. Gieseking (ed.). Springer-Verlag, New York, pp. 121-154.

Tan, K. H. (1964). The Andosols in Indonesia. In *F.A.O. meeting on the classification and correlation of soils from volcanic ash.* World Soil Resources, F.A.O. Reports No. 14, U.N. Educ. Sci. and Cultural Org., Rome, Italy, pp. 23-30.

Tan, K. H. (1975). Infrared absorption similarities between hymatomelanic acid and methylated humic acid. Soil Sci. Soc. Amer. Proc. *39*:70-73.

Tan, K. H. (1976a). Complex formation between humic acid and clays as revealed by gel filtration and infrared spectroscopy. Soil Biol. Biochem. *8*:235-239.

Tan, K. H. (1976b). Contamination of humic acid by silica gel and sodium bicarbonate. Plant and Soil *44*:691-695.

Tan, K. H. (1978a). Formation of metal-humic acid complexes by titration and their characterization by differential thermal analysis and infrared spectroscopy. Soil Biol. Biochem. *10*:123-129.

Tan, K. H. (1978b). Effect of humic and fulvic acids on release of fixed potassium. Geoderma *21*:67-74.

Tan, K. H. (1978c). Variations in soil humic compounds as related to regional and analytical differences. Soil Sci. *125*:351-358.

Tan, K. H., L. D. King, and H. D. Morris (1971a). Complex reactions of zinc with organic matter extracted from sewage sludge. Soil Sci. Soc. Amer. Proc. *35*:748-751.

Tan, K. H., R. A. Leonard, A. R. Bertrand, and S. R. Wilkinson (1971b). The metal complexing capacity and the nature of the chelating ligands of water extract of poultry litter. Soil Sci. Soc. Amer. Proc. *35*:266-269.

Tan, K. H., G. W. Bailey, and H. F. Perkins (1978). Infrared Analysis. In *Analysis of clay, silt and sand fractions of selected soils from the southeastern United States*, Southern Coop. series Bull. 219. University Kentucky, Lexington, Ky., Chapter 6, pp. 25-38.

Tan, K. H., E. R. Beaty, R. A. McCreery, and J. B. Jones (1975). Differential effect of bermuda and bahiagrasses on soil chemical characteristics. Agronomy J. *67*:407-411.

Tan, K. H., and F. E. Clark (1968). Polysaccharide constituents in fulvic and humic acids extracted in soils. Geoderma *2*:245-255.

Tan, K. H., and J. E. Giddens (1972). Molecular weights and spectral characteristics of humic and fulvic acids. Geoderma *8*:221-229.

Tan, K. H., and B. F. Hajek (1977). Thermal analysis of soils. In *Minerals and Soil Environments*, J. B. Dixon, S. B. Weed, J. A. Kittrick, M. H. Milford, and J. L. White (eds.). Soil Sci. Soc. Amer. Inc., Madison, Wis., pp. 865-884.

Tan, K. H., and R. A. McCreery (1975). Humic acid complex formation and intermicellar adsorption by bentonite. Proc. 1974 Intern. Clay Conference, Mexico City, Mexico. Applied, Wilmette, Ill., pp. 629-641.

Tan, K. H., and R. A. McCreery (1970a). Possibility of the silylation technique in gas liquid chromotography of fulvic and humic acids. Geoderma *4*:119-126.

Tan, K. H., and R. A. McCreery (1970b). The infrared indentification of a humo-polysaccharide ester in soil humic acid. Comm. Soil Sci. Plant Analysis *1*(2):75-84.

Tan, K. H., and V. Nopamornbodi (1979). Effect of different levels of humic acids on nutrient content and growth of corn (*Zea mays* L.). Plant and Soil *51*:283-387.

Tan, K. H., and H. F. Perkins (1980). The value of the U.S. Soil Taxonomy classification of red tropical soils derived from volcanic ash. Proc. CLAMATROPS, Intern. Conf. Classif. and Management of tropical soils, Kuala Lumpur, Malaysia, Aug. 1977. Intern. Soil Sci. Soc., Kuala Lumpur, Malaysia, pp. 110-119.

Tan, K. H., and J. Van Schuylenborgh (1959). On the classification and genesis of soils derived from andesitic volcanic material under a monsoon climate. Neth. J. Agric. Sci. *7*:1-22.

Tan, K. H., and J. Van Schuylenborgh (1961). On the organic matter in tropical soils. Neth. J. Agric. Sci. *9*:174-180.

Tan, K. H., V. G. Mudgal, and R. A. Leonard (1975). Adsorption of poultry litter extracts by soil and clay. Environm. Sci. and Techn. *9*:132-135.

Tan, K. H., H. F. Perkins, and R. A. McCreery (1973). Kaolinite-gibbsite thermodynamic relationship in Ultisols. Soil Sci. *116*:8-12.

Taylor, S. A., and G. L. Ashcroft (1972). *Physical Edaphology*. Freeman, San Francisco, Calif.

Taylor, S. A., and R. O. Slatyer (1960). Water-soil plant relations terminology. Trans. Intern. Congr. Soil Sci., Madison, 7th, *1*:80-90.

Theng, B. K. G. (1974). The chemistry of clay-organic reactions. Wiley, New York.

Theng, B. K. G. (1972). Formation, properties, and practical applications of clay-organic complexes. J. Royal Soc. New Zealand *2*:437-457.

Thomas, G. W. (1974). Chemical reactions controlling soil solution electrolyte concentration. In *Plant Root and Its Environment*, E. W. Carson (ed.). University Press of Virginia, Charlottesville, pp. 483-506.

Tisdale, S. L., and W. L. Nelson (1975). Soil fertility and fertilizers. Macmillan, New York.

Tiurin, I. V., and M. M. Kononova (1962). Biology of humus formation and questions of soil fertility. Trans. Joint Meeting Comm. IV and V, Int. Soc. Soil Sci., Palmerston-north, New Zealand, pp. 203-219.

USDA Soil Conservation Service, Soil Survey Staff (1960). Soil Classification, a comprehensive system, 7th approximation. U.S. Government Printing Office, Washington, D.C.

USDA Soil Conservation Service, Soil Survey Staff (1975). *Soil Taxonomy—A Basic System of Soil Classification for Making and Interpreting Soil Surveys*. Agriculture Handbook No. 436, USDA, SCS, U.S. Government Printing Office, Washington, D.C.

Van Breemen, N., and R. Brinkman (1976). Chemical equilibria and soil formation. In *Soil Chemistry, A. Basic Elements.* G. H. Bolt and M. G. M. Bruggenwert (eds.). Elsevier Sci., Amsterdam, pp. 141-170.

Van der Marel, H. W. (1959). Potassium fixation, a beneficial soil characteristic for crop production. Z. Pflanzenernährung, Dungung, Bodenkunde *84*(129):51-62.

Van Olphen, H. (1977). Clay colloid chemistry. Wiley, New York.

Van Raij, B., and M. Peech (1972). Electrochemical properties of some Oxisols and Alfisols of the tropics. Soil Sci. Soc. Amer. Proc. *36*:587-593.

Van Schuylenborgh, J. (1965). The formation of sesquioxides in soils. In *Experimental Pedology*, E. G. Hallsworth and D. V. Crawford (eds.). Butterworth, London, pp. 113-125.

Van Schuylenborgh, J. (1966). Chemical aspects of soil formation, syllabus of lectures postgraduate training. Agric. University Wageningen, Netherlands.

Van Schuylenborgh, J. (1971). Weathering and soil forming processes in the tropics. Proc. Bandung Symposium Soils and Trop. Weathering, Indonesia, Unesco, Paris, pp. 39-50.

Verwey, E. J. W., and J. T. G. Overbeek (1948). Theory of the stability of lyophobic colloids. Elsevier, New York

Wada, Koji (1977). Allophane and imogolite. In *Minerals in Soil Environments* J. B. Dixon, S. B. Weed, J. A. Kittrick, M. H. Milford, and J. L. White, (eds.). Soil Sci. Soc. Amer., Madison, Wis., pp. 603-639.

Walker, G. F. (1961). Vermiculite minerals. In *The X-ray identification and crystal structures of clay minerals.* G. Brown (ed.). Mineralogical Society, London, pp. 297-324.

Walker, G. F. (1975). Vermiculites. In *Soil Components*, Vol. 2, Inorganic Components, J. E. Gieseking (ed.). Springer-Verlag, New York, pp. 155-189.

Wallace, J. M., and L. C. Whitehand (1980). Adverse synergistic effects between acetic, propionic, butyric, and valeric acids on the growth of wheat seedling roots. Soil Biol. Biochem. *12*:445-446.

Way, J. T. (1850). On the power of soils to absorb manure. J. Roy. Agr. Soc. *11*:313-379.

Weast, R. C. (ed.) (1972). *Handbook of Chemistry and Physics.* The Chemical Rubber Co., Cleveland, Ohio.

Weber, J. B. (1970). Mechanisms of adsorption of s-triazines by clay colloids and factors affecting plant availability. In *Residue Reviews*, F. A. Gunther, and J. D. Gunther (eds.), Vol. 32. Springer-Verlag, New York, pp. 93-130.

White, J. L. (1971). Interpretation of infrared spectra of soil minerals. Soil Sci. *112*:22-31.

Yoshida, M., K. Sakagami, R. Hamada, and T. Kurobe (1978). Studies on the properties of organic matter in buried humic horizon derived from volcanic ash. I. Humus composition of buried humic horizon. Soil Sci. Plant Nutr. *24*:277-287.

Yoshinaga, N., and S. Aomine (1962). Imogolite in some Ando soils. Soil Sci. Plant Nutr. *8*:6-13.

AUTHOR INDEX

Italic numbers give the page on which the complete reference is listed.

SUBJECT INDEX

PIETÀ

★

R. S. THOMAS

RUPERT HART-DAVIS
SOHO SQUARE LONDON
1966

First Published 1966
Rupert Hart-Davis Ltd
36 Soho Square London W1

Printed in Great Britain by
Cox & Wyman Ltd
London, Fakenham and Reading

CONTENTS

RHODRI

Rhodri Theophilus Owen,
Nothing Welsh but the name;
He moves in a landscape of dust
That is sourer than the smell
Of breweries. What are the moors
To him? Shadows of boredom
In the mind's corners. He has six shirts
For the week-end and a pocketful
Of notes. Don't mention roots
To Rhodri; his address
Is greater than the population
Of Dolfor, many times
Greater, and in that house
There are three Owens, none with a taste
For the homeland with its pints
Of rain water.
It is dry
Here, with the hard, dry
Urban heat, that is sickly
With girls. But Rhodri is cool;
From the shadow of his tree
Of manhood he watches them
Pass, or selects one
To make real the power of the pounds,
That in Wales would have gone rather
To patch up the family stocking,
Emblem of a nation's despair.

BECAUSE

I praise you because
I envy your ability to
See these things: the blind hands
Of the aged combing sunlight
For pity; the starved fox and
The obese pet; the way the world
Digests itself and the thin flame
Scours. The youth enters
The brothel, and the girl enters
The nunnery, and a bell tolls.
Viruses invade the blood.
On the smudged empires the dust
Lies and in the libraries
Of the poets. The flowers wither
On love's grave. This is what
Life is, and on it your eye
Sets tearless, and the dark
Is dear to you as the light.

SWIFTS

The swifts winnow the air.
It is pleasant at the end of the day
To watch them. I have shut the mind
On fools. The phone's frenzy
Is over. There is only the swifts'
Restlessness in the sky
And their shrill squealing.
Sometimes they glide,
Or rip the silk of the wind
In passing. Unseen ribbons
Are trailing upon the air.
There is no solving the problem
They pose, that had millions of years
Behind it, when the first thinker
Looked at them.
Sometimes they meet
In the high air; what is engendered
At contact? I am learning to bring
Only my wonder to the contemplation
Of the geometry of their dark wings.

ROSE COTTAGE

Rose Cottage, because it had
Roses. If all things were as
Simple! There was the place
With some score or so of
Houses, all of them red
Brick, with their names clear
To read; and this one, its gate
Mossed over, its roof rusty
With lichen. You chose it out
For its roses, and were not wrong.
It was registered in the heart
Of a nation, and so, sure
Of its being. All summer
It generated the warmth
Of its blooms, red lamps
To guide you. And if you came
Too late in the bleak cold
Of winter, there were the faces
At the window, English faces
With red cheeks, countering the thorns.

HAFOD LOM

Hafod Lom, the poor holding:
I have become used to its
Beauty, the ornamentation
Of its bare walls with grey
And gold lichen; to its chimney
Tasselled with grasses. Outside
In the ruined orchard the leaves
Are richer than fruit; music
From a solitary robin plays
Like a small fountain. It is hard
To recall here the drabness
Of past lives, who wore their days
Raggedly, seeking meaning
In a lean rib. Imagine a child's
Upbringing, who took for truth
That rough acreage the rain
Fenced; who sowed his dreams
Hopelessly in the wind blowing
Off bare plates. Yet often from such
Those men came, who, through windows
In the thick mist peering down
To the low country, saw learning
Ready to reap. Their long gnawing
At life's crust gave them teeth
And a strong jaw and perseverance
For the mastication of the fact.

THIS TO DO

I have this that I must do
One day: overdraw on my balance
Of air, and breaking the surface
Of water go down into the green
Darkness to search for the door
To myself in dumbness and blindness
And uproar of scared blood
At the eardrums. There are no signposts
There but bones of the dead
Conger, no light but the pale
Phosphorous, where the slow corpses
Swag. I must go down with the poor
Purse of my body and buy courage,
Paying for it with the coins of my breath.

WITHIN SOUND OF THE SEA

I have a desire to walk on the shore,
To visit the caged beast whose murmurings
Kept me awake. What does it mean
That I have the power to do this
All day long, if I wish to?
I know what thoughts will arise,
What questions. They have done so before,
Unanswered. It is in the freedom
To go or not to I exist;
To balance all the exhilaration
Of brisk moments upon the sand
With the knowledgeable hours that my books
Give me. Between their pages
The beast sleeps and never looks out
Through the print's bars. Have I been wise
In the past, letting my nostrils
Plan my day? That salt scrubbing
Left me unclean. Am I wise now,
With all this pain in the air,
To keep my room, reading perhaps
Of that Being whose will is our peace?

PIETÀ

Always the same hills
Crowd the horizon,
Remote witnesses
Of the still scene.

And in the foreground
The tall Cross,
Sombre, untenanted,
Aches for the Body
That is back in the cradle
Of a maid's arms.

AMEN

And God said: How do you know?
And I went out into the fields
At morning and it was true.

Nothing denied it, neither the bowed man
On his knees, nor the animals,
Nor the birds notched on the sky's

Surface. His heart was broken
Far back, and the beasts yawned
Their boredom. Under the song

Of the larks, I heard the wheels turn
Rustily. But the scene held;
The cold landscape returned my stare;

There was no answer. Accept; accept.
And under the green capitals,
The molecules and the blood's virus.

THE PROVINCIAL

He is that dark side
Of you that you keep
Hidden, the poor relation
You avoid. Since you left
The villages' and the fields'
Mire, you have looked
Forward only at your reflection
In plastics. Do you never
Pause thoughtful before the trash
Of windows, and on scenes
Of false snow see bleakness
Of the real world impose
His figure, who is native
Of such truth, and makes his way
With dignity to the same
Poor, cold, bare
Resting place as do you
With your immense solvency for show?

GIFTS

From my father my strong heart,
My weak stomach.
From my mother the fear.

From my sad country the shame.

To my wife all I have
Saving only the love
That is not mine to give.

To my one son the hunger.

And beyond the window Denmark
Waited, but refused to adopt
This family that wore itself out
On its conscience, up and down
In the one room.
                    Meanwhile the acres
Of the imagination grew
Unhindered, though always they paused
At that labourer, the indictment
Of whose gesture was a warped
Crucifix upon a hill
In Jutland. The stern father
Looked at it and a hard tear
Formed, that the child's frightened
Sympathy could not convert
To a plaything.
                    He lived on,
Søren, with the deed's terrible lightning
About him, as though a bone
Had broken in the adored body
Of his God. The streets emptied
Of their people but for a girl
Already beginning to feel
The iron in her answering his magnet's
Pull. Her hair was to be
The moonlight towards which he leaned
From darkness. The husband stared
Through life's bars, venturing a hand
To pluck her from the shrill fire
Of his genius. The press sharpened

Its rapier; wounded, he crawled
To the monastery of his chaste thought
To offer up his crumpled amen.

FOR INSTANCE

She gave me good food;
I accepted;

Sewed my clothes, buttons;
I was smart.

She warmed my bed;
Out of it my son stepped.

She was adjudged
Beautiful. I had grown

Used to it. She is dead
Now. Is it true

I loved her? That is how
I saw things. But not she.

RAVENS

It was the time of the election.
The ravens loitered above the hill
In slow circles; they had all air
To themselves. No eyes were lifted
From the streets, no ears heard
Them exulting, recalling their long
History, presidents of the battles
Of flesh, the sly connoisseurs
Of carrion; desultory flags
Of darkness, saddening the sky
At Catraeth and further back,
When two, who should have been friends,
Contended in the innocent light
For the woman in her downpour of hair.

FOR THE RECORD

What was your war record, Prytherch?
I know: up and down the same field,
Following a horse; no oil for tractors;
Sniped at by rain, but never starving.
Did you listen to the reports
Of how heroes are fashioned and how killed?
Did you wait up late for the news?
Your world was the same world as before
Wars were contested, noisier only
Because of the echoes in the sky.
The blast worried your hair on its way to the hill;
The distances were a wound
Opened each night. Yet in your acres,
With no medals to be won,
You were on the old side of life,
Helping it in through the dark door
Of earth and beast, quietly repairing
The rents of history with your hands.

A WELSHMAN AT ST. JAMES' PARK

I am invited to enter these gardens
As one of the public, and to conduct myself
In accordance with the regulations;
To keep off the grass and sample flowers
Without touching them; to admire birds
That have been seduced from wildness by
Bread they are pelted with.

I am not one
Of the public; I have come a long way
To realise it. Under the sun's
Feathers are the sinews of stone,
The curved claws.

I think of a Welsh hill
That is without fencing, and the men,
Bosworth blind, who left the heather
And the high pastures of the heart. I fumble
In the pocket's emptiness; my ticket
Was in two pieces. I kept half.

THE MOOR

It was like a church to me.
I entered it on soft foot,
Breath held like a cap in the hand.
It was quiet.
What God was there made himself felt,
Not listened to, in clean colours
That brought a moistening of the eye,
In movement of the wind over grass.

There were no prayers said. But stillness
Of the heart's passions—that was praise
Enough; and the mind's cession
Of its kingdom. I walked on,
Simple and poor, while the air crumbled
And broke on me generously as bread.

GIRL

And her breath that is like
A vase of flowers, of dead flowers. I take
Her hand with its red nails
In mine, and examine her eyelids'
Mascara, and what is left
Of her brows. And her teeth manage
Some laughter that breaks its china
Upon me.
Ageless girl,
With your propped charm and your sex's
Tinkering with the wind
At the nostril, need you care
What I think? Half the world
Hesitates at its dull prayers,
As its soul skids suddenly on your stocking.

THERE

They are those that life happens to.
They didn't ask to be born
In those bleak farmsteads, but neither
Did they ask not. Life took the seed
And broadcast it upon the poor,
Rush-stricken soil, an experiment
In patience.

What is a man's
Price? For promises of a break
In the clouds; for harvests that are not all
Wasted; for one animal born
Healthy, where seven have died,
He will kneel down and give thanks
In a chapel whose stones are wrenched
From the moorland.

I have watched them bent
For hours over their trade,
Speechless, and have held my tongue
From its question. It was not my part
To show them, like a meddler from the town,
Their picture, nor the audiences
That look at them in pity or pride.

ON TOUR

And the towns say:
Spare us some of your hate.
And the signs lie for them:
What saint would reside here?
Their breath has become smoke.
Irritable summers
Smear the pavements with sun,
The trees are rusty.

I have grown used
To the accusation of the hills.
There is no answer;
This people cannot translate
Beauty. I must forgive them.
They sin in Welsh.

THE BELFRY

I have seen it standing up grey,
Gaunt, as though no sunlight
Could ever thaw out the music
Of its great bell; terrible
In its own way, for religion
Is like that. There are times
When a black frost is upon
One's whole being, and the heart
In its bone belfry hangs and is dumb.

But who is to know? Always,
Even in winter in the cold
Of a stone church, on his knees
Someone is praying, whose prayers fall
Steadily through the hard spell
Of weather that is between God
And himself. Perhaps they are warm rain
That brings the sun and afterwards flowers
On the raw graves and throbbing of bells.

ASIDE

Take heart, Prytherch.
Over you the planets stand,
And have seen more ills than yours.
This canker was in the bone
Before man bent to his image
In the pool's glass. Violence has been
And will be again. Between better
And worse is no bad place

For a labourer, whose lot is to seem
Stationary in traffic so fast.
Turn aside, I said; do not turn back.
There is no forward and no back
In the fields, only the year's two
Solstices, and patience between.

THE VISIT

She was small;
Composed in her way
Like music. She sat
In the chair I had not
Offered, smiling at my left
Shoulder. I waited on
For the sentences her smile
Sugared.

That the tongue
Is a whip needed no
Proving. And yet her eye
Fondled me. It was clear
What anger brought her
To my door would not unleash
The coils. Instead she began
Rehearsing for her
Departure. As though ashamed
Of a long stay, she rose,
Touched the tips of my cold
Hand with hers and turned
To the closed door. I remember
Not opening it.

EXCHANGE

She goes out.
I stay in.
Now we have been
So long together
There's no need
To share silence;
The old bed
Remains made
For two; spirits
Mate apart
From the sad flesh,
Growing thinner
On time's diet
Of bile and gall.

GOSPEL TRUTH

MARK PUW: Who put me here?
I must get on with this job;
The rain is coming. I must get on
With this job. Who put me
Here? The bugger; I'd like
To see him now in my place.

MATTHEW PUW: There are times
When I could wreck the whole bloody
Farm. Pig music, sheep music, the grey
Traffic of the clouds going by
I could get shut of the lot,
If it wasn't for him and Mair.

LUKE PUW: Every morning I wake,
There is the spilled light in my room;
I rise and try to wash my hands
In its cold water. Then inside
The tune starts; the rest leave me
To play it over to myself.

MAIR PUW: There are the clothes
To wash Mondays, and the plates
To be kept filled; boots to clean.
I have stopped trying to think
What I should do, if that face
In the half-darkness should go out.

MATTHEW PUW: Heaven and heaven's foundations
Rot. I have seen the girl,

Whose flesh is like the warm milk
In pails I have carried these ten years
Without tasting. I am thirsty
As the hard earth in spring and as dry.

MARK PUW: I shall get all the blame
For this. Who put the seed down
In such places? I never wanted
The ground ploughed; better to have left it
As earth wanted it to be,
Fertile of stone, the wind's pasture.

LUKE PUW: Leaves of the sun
Are falling; it is always autumn
Where I go. On bare branches
I look into the eyes of the dumb
Blackbirds. What I sing
Frightens them; it is called 'Luke Puw'.

MAIR PUW: I have watched them come
In from the fields more times than I wish
To be told. My heart is the clapper
Of an old bell, stiff and rusty.
How long will those hands pull
On the rope fastening it to care?

MATTHEW PUW: Under the talk
I am listening to the dusk's
Sounds; the owl and badger
Waken. But that familiar
Drooling has not begun yet
To drill the nerve. Is this it?

MARK PUW: There is an empty
Place at the table; instead
Of a respite from that tune
In the skull, I must endure
The accusation of dead eyes
From a portrait. It is not my fault.

MAIR PUW: This has happened
Before. If I bring myself
To stay still, they will go out
Like dogs that round up a strayed
Sheep. But that inaudible
Whistle—from whose lips does it come?

THEN

It was one of those days not confined
To a country, most of the earth
Knows them: precipiced clouds
Sheer above the blue
Chasms, the sun not hot
But registering for the first time
Its presence, opening the heart's
Flower that winter had clenched.

We wandered upon the broad hills'
Back, crumbling the air's
Poetry. Nothing that nature
Did was a contradiction
That time, and the prey hung
Jewels of blood round the day's throat.

SERVICE

We stand looking at
Each other. I take the word 'prayer'
And present it to them. I wait idly,
Wondering what their lips will
Make of it. But they hand back
Such presents. I am left alone
With no echoes to the amen
I dreamed of. I am saved by music
From the emptiness of this place
Of despair. As the melody rises
From nothing, their mouths take up the tune,
And the roof listens. I call on God
In the after silence, and my shadow
Wrestles with him upon a wall
Of plaster, that has all the nation's
Hardness in it. They see me thrown
Without movement of their oblique eyes.

BLONDES

They pass me with bland looks.
It is the simplicity of their lives
I ache for: prettiness and a soft heart, no problems
Not to be brought to life size
By a kiss or a smile. I see them walking
Up long streets with the accuracy of shuttles
At work, threads crossed to make a pattern
Unknown to them. A thousand curtains
Are parted to welcome home
The husbands who have overdrawn
On their blank trust, giving them children
To play with, a jingle of small change
For their pangs. The tear-laden tree
Of a poet strikes no roots in their hearts.

THE DANCE

She is young. Have I the right
Even to name her? Child,
It is not love I offer
Your quick limbs, your eyes;
Only the barren homage
Of an old man whom time
Crucifies. Take my hand
A moment in the dance,
Ignoring its sly pressure,
The dry rut of age,
And lead me under the boughs
Of innocence. Let me smell
My youth again in your hair.

WHO?

Someone must have thought of putting me here;
It wasn't myself did it.
What do I find to my taste?
Annually the grass comes up green;
The earth keeps its rotary motion.
There is loveliness growing, where might have been truth's
Bitterer berries. The reason tempers
Most of the heart's stormier moods.

But there's an underlying despair
Of what should be most certain in my life:
This hard image that is reflected
In mirrors and in the eyes of my friends.
It is for this that the air comes in thin
At the nostril, and dries to a crust.

BROTHERS

Dewi is out paddling in the grass.
Ianto, he is doing the same.
Llywelyn John? He was here just now.

Don't go out on the hill's brow
To look for them. Let them lie
Safe and sane in their names' cradle.

Let the round mind fill their place
At the scrubbed table with tall forms,
Beautiful in the half-light.

Let the old mother rambling on
Interminably as the wind keep them
For you, as for herself, young.

It is best so. What have those shadows,
Torturing the fields' peace
Like dead trees, to do with a man?

Slowly what should have been a face
Turns itself to the bare hill
And throws up the remains of laughter.

THE FACE

When I close my eyes, I can see it,
That bare hill with the man ploughing,
Corrugating that brown roof
Under a hard sky. Under him is the farm,
Anchored in its grass harbour;
And below that the valley
Sheltering its few folk,
With the school and the inn and the church,
The beginning, middle and end
Of their slow journey above ground.

He is never absent, but like a slave
Answers to the mind's bidding,
Endlessly ploughing, as though autumn
Were the one season he knew.
Sometimes he pauses to look down
To the grey farmhouse, but no signals
Cheer him; there is no applause
For his long wrestling with the angel
Of no name. I can see his eye
That expects nothing, that has the rain's
Colourlessness. His hands are broken
But not his spirit. He is like bark
Weathering on the tree of his kind.

He will go on; that much is certain.
Beneath him tenancies of the fields
Will change; machinery turn
All to noise. But on the walls

Of the mind's gallery that face
With the hills framing it will hang
Unglorified, but stern like the soil.

SCHOONERMEN

Great in this,
They made small ships do
Big things, leaping hurdles
Of the stiff sea, horse against horses
In the tide race.
                    What has Rio
To do with Pwllheli? Ask winds
Bitter for ever
With their black shag. Ask the quays
Stained with spittle.
                    Four days out
With bad cargo
Fever took the crew;
The mate and boatswain,
Peering in turn
Through the spray's window,
Brought her home. Memory aches
In the bones' rigging. If tales were tall,
Waves were taller.
                    From long years
In a salt school, caned by brine,
They came landward
With the eyes of boys,
The Welsh accent
Thick in their sails.

IN CHURCH

Often I try
To analyse the quality
Of its silences. Is this where God hides
From my searching? I have stopped to listen,
After the few people have gone,
To the air recomposing itself
For vigil. It has waited like this
Since the stones grouped themselves about it.
These are the hard ribs
Of a body that our prayers have failed
To animate. Shadows advance
From their corners to take possession
Of places the light held
For an hour. The bats resume
Their business. The uneasiness of the pews
Ceases. There is no other sound
In the darkness but the sound of a man
Breathing, testing his faith
On emptiness, nailing his questions
One by one to an untenanted cross.

AH!

There's no getting round it,
It's a hell of a thing, he said, and looked grave
To prove it. What he said was
The truth. I would make different
Provision; for such flesh arrange
Exits down less fiery paths. But the God
We worship fashions the world
From such torment, and every creature
Decorates it with its tribute of blood.